Essentials of the Earth's Climate System

This concise, full-color introduction to modern climatology covers all the key topics of climate science for undergraduate/graduate students on one-semester courses. The book progresses from climate processes to world climate types, past climate change and projected future climates, ending with climate applications. The treatment of topics is non-mathematical wherever possible, allowing students to understand the physical processes more easily.

- Clear, full-color illustrations support the topics introduced in the text.
- Boxes are used to provide supplementary topics, enabling students to increase their knowledge and awareness.
- Each chapter concludes with a summary of the main points and a mixture of review and discussion questions that encourage students to check their understanding and think critically.

Roger G. Barry was born in Sheffield, England and worked for two years in the British Meteorological Office before attending Liverpool University, where he received a BA Honors in Geography. He received an MSc in Climatology from McGill University, Montreal, and a PhD from the University of Southampton in 1965. In October 1968 he moved to the University of Colorado, Boulder to become Associate Professor of Geography, Professor (1971–2004), and Distinguished Professor (2004–2010). In 1977 he became the Director of the World Data Center for Glaciology, which in 1980 merged into the National Snow and Ice Data Center (NSIDC). Roger's teaching and research has spanned climate change, Arctic and mountain climates, synoptic climatology, and snow and ice processes. Roger has published 25 textbooks, including: *Atmosphere, Weather and Climate* (with R. J. Chorley, tenth edition, 2010); *Mountain Weather and Climate* (third edition, 2008, Cambridge University Press); *Synoptic and Dynamic Climatology* (with A. M. Carleton, 2011, Routledge); *The Arctic Climate System* (with M. C. Serreze, 2005, Cambridge University Press); *The Global Cryosphere: Past, Present and Future* (with T. Y. Gan, 2011, Cambridge University Press). He has also published more than 250 research articles, and supervised 65 graduate students. Roger has been a Guggenheim Fellow, a Fulbright Teaching Fellow at Moscow State University, and a Visiting Professor in Australia, France, Germany, Japan, New Zealand, Switzerland, and the United Kingdom. His honors include: Fellow, American Geophysical Union; Foreign Member of the Russian Academy of Environmental Science (RAEN) Founder's Medal, Royal Geographical Society; Humboldt Prize Fellow. He is currently Director of the International CLIVAR Project Office at the National Oceanography Centre, Southampton, UK.

Eileen A. Hall-McKim is a PhD Climatologist receiving her degree from the University of Colorado, Boulder. Her interdisciplinary degrees include work in the geological sciences, hydrology, oceanography, paleoclimatology, and water resource research. She completed her MSc at the National Snow and Ice Data Center, Boulder and worked as editor and writer for the Intermountain West Climate Summary of the NOAA/Western Water Assessment. Her honors include: elected member Phi Beta Kappa National Honor Society; Outstanding Women in Geosciences Student Award from the American Association of Women in Geosciences; Graduate Research Fellowship Award from the Cooperative Institute for Research in Environmental Sciences (CIRES); Magna Cum Laude National Honor Society, University of Colorado. She is currently pursuing professional certification in sustainable practices from the University of Colorado law school and environmental science department.

D1334690

"Absolutely the ultimate word on the physical, synoptic, and geographic underpinnings of modern climatology and its historic antecedents. This delightful and readable textbook covers all the topics – and then some – likely to comprise an introductory or intermediate-level college course . . . Discussions of the forcing, form, and function of the climate system – on scales ranging from local to global – will be essential reading for university undergraduate and graduate students alike, and a helpful review for seasoned researchers in the climate and atmospheric sciences. The easy-on-the-eye text style is complemented by the many incisive color figures, maps, and graphs, most of which are based on the latest analyses from satellites and global re-analysis data. Beginning and end-of-chapter overviews and summaries highlight the most important concepts and features of climatology study, while the glossary and bibliography are both comprehensive and fully up to date. *Essentials of the Earth's Climate System* is the essential text on climatology, and is destined to become a classic."

Andrew M. Carleton, *Pennsylvania State University*

"This textbook is a very comprehensive and informative resource for teaching and for general reference. Its layout and organization are efficient and effective, allowing a wide range of material to be covered in a surprising level of detail . . . This book has an important place in the classroom and on any Earth scientist's bookshelf."

David A. Pepper, *California State University, Long Beach*

"This textbook provides a comprehensive and well-illustrated overview of the climate system by experts with a wealth of experience in climate science."

Raymond S. Bradley, *University of Massachusetts*

"I can recommend this text, particularly for students studying an introduction to climate science at undergraduate level. The text is accessible and any mathematical treatment is clearly explained and at an introductory level. I am particularly impressed by the scope of material, with chapters on past climates, future climate modeling, and applied climatology, a welcome addition to the usual material on atmospheric systems and local/regional climates. It is also great to see case studies illustrated with examples from all over the world. This book will be a comprehensive resource for all those teaching climate science at an introductory level."

Nicholas Pepin, *University of Portsmouth*

" . . . an excellent introduction to climate science, enabling coverage of the main issues in one semester and an inspiration for more in-depth studies. It is simple enough to be understood by geography and environmental science undergraduates without previous knowledge of climatology, but not oversimplified. It skillfully mixes complex concepts with observational data engaging the reader and making the challenging understandable and the seemingly tedious exciting . . . An immense advantage of this textbook is that its first part provides excellent explanations of the very basics of climate science . . . The authors manage to lay the foundations for more advanced studies and engage readers through the use of diverse examples from various parts of the world . . . The authors skillfully intermingle observational data with explanations of complex processes and concepts in an engaging and easy-to-follow manner . . . Overall, I highly recommend this book for undergraduate courses, and every university library should have a copy."

Maria Shahgedanova, *University of Reading*

"Drawing on more than 50 years of combined experience in climate study, Barry and Hall-McKim give the reader a compact, non-mathematical overview of the fundamental processes of the Earth's climate system. The book is surprisingly comprehensive, given its relatively brief length: coverage ranges from global to the local, from short-term phenomena to long-term climatic change. Complex topics are explained in straightforward, non-complex language, which in turn is supported by excellent color illustrations, and numerous place-specific examples are skillfully employed to illustrate general processes and concepts. *Essentials of the Earth's Climate System* is an ideal introduction to the topic for an upper-level undergraduate course in climate. It is likely to become the standard textbook in its field."

Thomas Krabacher, *California State University, Sacramento*

Essentials of the Earth's Climate System

Roger G. Barry

Eileen A. Hall-McKim

UNIVERSITY OF COLORADO AT BOULDER

CAMBRIDGE
UNIVERSITY PRESS

CAMBRIDGE
UNIVERSITY PRESS

University Printing House, Cambridge CB2 8RU, United Kingdom

Published in the United States of America by Cambridge University Press, New York

Cambridge University Press is part of the University of Cambridge.

It furthers the University's mission by disseminating knowledge in the pursuit of education, learning, and research at the highest international levels of excellence.

www.cambridge.org
Information on this title: www.cambridge.org/9781107037250

© Roger G. Barry and Eileen A. Hall-McKim 2014

First published 2014

Printed in Singapore by C.O.S. Printers Pte Ltd

A catalogue record for this publication is available from the British Library

Library of Congress Cataloguing in Publication data
Barry, Roger Graham.
Essentials of the earth's climate system / Roger Barry, Eileen A. Hall-McKim.
    pages   cm
ISBN 978-1-107-03725-0 (Hardback)
1. Climatology – Textbooks.   I. Hall-McKim, Eileen A.   II. Title.
QC981.B3265   2013
551.6–dc23   2013022002

ISBN 978-1-107-03725-0 Hardback
ISBN 978-1-107-62049-0 Paperback

Additional resources for this publication at www.cambridge.org/climatesystem

Contents

Preface   *page* ix
Acknowledgments   xi

1   Introduction   1
    1.1   Climate and weather   2
    1.2   Why climate matters   3
    1.3   Climate statistics   4
    1.4   History of world climatology   6
    1.5   The use of weather satellites   11

2   The elements of climate: a global view of energy and
     moisture   14
    2.1   Energy   15
           (a) Solar radiation   15
           (b) Cloud effects   18
           (c) Aerosol effects   19
           (d) Surface receipt of solar radiation   20
           (e) Infrared radiation   21
           (f) Net radiation   23
           (g) Heat fluxes   23
           (h) Temperature   25
    2.2   Moisture   29
           (a) Vapor content, vapor pressure, relative humidity, and precipitable
                water   29
           (b) Clouds   31
           (c) Lightning   32
           (d) Precipitation   34
           (e) Evaporation   40
           (f) Visibility   45

3   The elements of climate: a global view of pressure,
     winds, and storms   47
    3.1   Pressure and winds   48
           (a) Pressure   48
           (b) Winds   50
           (c) Horizontal moisture flux   57
    3.2   Air masses   59
    3.3   Frontal zones   61

3.4	Storm frequency and tracks	64
3.5	Thunderstorms	71
3.6	Mesoscale convective systems	71

4	**Local and microclimates**	**74**
4.1	Local climate	75
4.2	Forest climate	79
4.3	Lake climate	80
4.4	Urban climate	81
	(a) Pollution	82
	(b) Urban heat island	82
	(c) Moisture effects	84
4.5	Microclimate	85
4.6	Human bioclimatology	86

5	**The general circulation**	**91**
5.1	Factors	92
5.2	Meridional cells and zonal winds	94
5.3	Zonal wind belts	96
5.4	Zonal circulations	100

6	**Circulation modes**	**103**
6.1	Introduction	103
6.2	The Southern Oscillation and El Niño	104
6.3	The North Atlantic Oscillation	109
6.4	The Northern Annular Mode	109
6.5	The Southern Annular Mode	110
6.6	The semiannual oscillation	110
6.7	The North Pacific Oscillation	110
6.8	The Pacific Decadal Oscillation	111
6.9	The Atlantic Multidecadal Oscillation	111
6.10	Southern Hemisphere wave number-three pattern	112
6.11	The Madden–Julian Oscillation	112

7	**Synoptic climatology**	**116**
7.1	Introduction	117
7.2	Regional classifications	117
7.3	Continental classifications	122
7.4	Hemispheric classifications	122
7.5	Modern applications of synoptic climatology	125

8   Land and sea effects   127
  8.1   Oceans   128
    (a) Northern Hemisphere currents   129
    (b) The Arctic Ocean   132
    (c) Southern Hemisphere currents   134
    (d) The Southern Ocean   134
  8.2   Air–sea interaction   135
    (a) Tropical cyclones   138
  8.3   Land   138

9   Climatic types on land   145
  9.1   Classifying climates   146
  9.2   Major climatic types on land   146
    (a) Deserts   146
    (b) Monsoons   149
    (c) High plateaus   154
    (d) Wet lowlands   157
    (e) Tropical and subtropical steppe   160
    (f) Humid subtropical   160
    (g) Mediterranean   161
    (h) Temperate lowlands   162
    (i) Maritime west coasts   164
    (j) Mid-latitude steppe and prairie   165
    (k) Taiga/boreal forest   165
    (l) Tundra   167
    (m) Ice plateaus   168

10   Past climates   172
  10.1   Geologic time   173
  10.2   The Cenozoic   175
  10.3   The Quaternary   176
  10.4   The Anthropocene   186

11   Future climate   194
  11.1   Global climate models   195
  11.2   Projected changes   196
  11.3   Impacts   199
  11.4   Economic and socio-political issues   202

12   Applied climatology   206
  12.1   Climatic extremes and disasters   207
  12.2   Climatic aspects of vegetation and soils   209

12.3   Agriculture and climate   212
12.4   Water resources   214
12.5   Renewable energy   217
12.6   Climate effects on transportation   219
12.7   Insurance and climate/weather disasters   220
12.8   Climate forecasts and climate services   221

Appendix A: Units   225
Appendix B: Web links   226
Glossary   228
Bibliography   245
Index   255

Preface

This textbook seeks to provide a modern global overview of the world's climates on all space and time scales. It addresses microclimates to global scale processes and phenomena. It spans climate changes over geologic time and the future climates of the late twenty-first century. It is designed to serve as an introductory course in climatology, suitable for students in environmental sciences, geography, meteorology, and related disciplines. The purpose of the book is first to provide a firm foundation of the physical principles that underpin climatology; second, to describe the spatial climatic characteristics over the globe including local and microclimatic scales; third, to detail the past and projected future climates of the Earth; and fourth, to introduce some applications of climatic information.

The book is organized into 11 parts following a brief introduction on definitions, statistics, and the history of climatology. These are: a global view of the major climatic elements of energy and moisture followed by pressure, wind and storms, local and microclimates, the general circulation, circulation modes, synoptic climatology, the regional effects of land and sea, climatic types on land, past climates, future climate and its impacts, and different examples of applied climatology. Chapters 2, 3, 5, 6, and 7 are more meteorological in content. Chapters 8 and 9 provide detailed accounts of oceanic and land climates.

My meteorological experience began in the early 1950s, when I worked as a scientific assistant in the British Meteorological Office for two years at Royal Air Force (RAF) station Worksop in Nottinghamshire and then, following an undergraduate degree at the University of Liverpool, in 1957–1958 I was a graduate student weather observer at the McGill Subarctic Research Station at Schefferville in northern Quebec-Labrador.

I have carried out meteorological fieldwork in the Canadian Arctic, Papua New Guinea, the Colorado Rocky Mountains, and the Venezuelan Andes. Among the climatologists featured in text boxes, I have personally known Hubert Lamb, Jerry Namias, Murray Mitchell, Herman Flohn, and Herbert Riehl.

The text builds on over 50 years' experience in teaching climatology to geography students at the University of Southampton, UK (1960–1968) and the University of Colorado, Boulder, Colorado, USA (1968–2010).

This textbook contains many **pedagogical features**:

- The treatment is non-mathematical, but physical processes are explained. Where simple equations are introduced, their meaning is fully explained.
- Clear illustrations in full colour support the topics introduced in the text.
- Two types of boxes are used: 'A' boxes for elaborations of points made in the text; and 'B' boxes for wider topics that can be used by teachers and students to expand their information and awareness.
- The main points of each chapter are recapped in a summary section at the end of chapters.
- A mixture of review and discussion questions encourage students to check their understanding and think critically.
- Students are further supported by a glossary, a list of useful websites, and an appendix on units.

The book is supported by a number of **online resources**, to be found at www.cambridge.org/climatesystem:

- Web links to data and other key resources.
- Solutions and hints to answers to the student questions (password-protected for access by course instructors).
- PowerPoint slides and JPEGS of all the figures in the book for the use of course instructors.

Roger G. Barry
Distinguished Professor of Geography Emeritus
Director, NSIDC/WDC for Glaciology, 1976–2008,
Director, International CLIVAR Project Office,
National Oceanography Centre, NERC, Southampton

Acknowledgments

We would like to thank Dr. Andrew Carleton for reviewing a draft of this book and making valuable suggestions. We also thank Sam Massey, Cody Sanford, Will Harris, Yana Duday, Amy Randall, and Jill Rittersbacher in the NSIDC Message Center for their help with diagrams.

We are grateful to the following for permission to reproduce figures:

American Geographical Society
Soviet Geography, 1962, 3(5) p.7, Fig. 2 and p. 9, Fig. 3
M. I. Budyko

American Geophysical Union
Rev. Geophys., 1999, 37: p. 193, Fig. 4; 2012, 50: RG4003, p. 20 Fig. 11
Geophys. Res. Lett., 2011, 38: L21218, Fig. 1
J. Geophys. Res., 2002, 116: D11108, Fig. 3

American Meteorological Society
Bull. Amer. Meteor. Soc. 2005, 86: 1456, Fig. 1
J. Climate, 1993, 6: p. 2168, Fig. 4; 2008, 21: p. 2322, Fig. 7d
Mon. Weather. Rev., 2004, 132: p. 1743, Fig. 23
Frederick Sanders Symposium, American Meteorological Society, 12 January
    2004, Seattle, WA, Paper 16.3, Figures 1 and 2.

Cambridge University Press, *Climate Change 2007: The Physical Science Basis.*
    Contribution of Working Group I to the Fourth Assessment Report of the
    Intergovernmental Panel on Climate Change (Eds. Solomon, S. and Qin, D. H.
    et al.) p. 767, Fig. 10.9

Colorado State University Department of Atmospheric Sciences, for Fig. 3.6.

Katie Davidson, National Oceanography Centre, Southampton for drafting
    Fig. 2.6

European Geophysical Union, *Atmos. Chem. Phys. Discuss.* 5, 2005, 4591, Fig. 2

European Space Agency
www.globvapour.info, User Help Desk

Food and Agriculture Organization, FAO Fisheries Technical Paper. No. 410.
    2001. Rome. L. B. Klyashtorin, p. 7, Fig. 2.2

C. Godske, 1966. "Methods of statistics and some applications to climatology," in
    Statistical Analysis and Prognosis in Meteorology. Tech, Note No. 71 (WMO

No, 178, TP 88), Geneva: World Meteorological Organization, p. 20, Figure 5.

Dr. P. D. Jones, Climatic Research Unit, University of East Anglia, for providing the data for Fig. 2.6

Dr. J. Kimball, University of Montana, for AMSR-E data on water surfaces.

Michael E. Mann, Pennsylvania State University for Fig. 10.5.

NASA Earth Observatory

Routledge/Taylor and Francis

R. G. Barry and R. J. Chorley, 2010, *Atmosphere, Weather and Climate*, 9th edn., p. 87, Fig. 4.6; p. 156 Parts B and D of Figure 6.11; p. 166, Fig. 7.3; p. 167, Fig. 7.4; p. 245, Fig. 9.18; p. 247, Fig. 9.21; p. 337, Fig. 11.8; p. 349, Fig. 11.8; p. 353, Fig. 11.23.

R. G. Barry in R. J. Chorley and P. Haggett, "Models in Geography," 1967, p.101, Figs. 4.1, 4.2, 4.3; p. 115, Fig. 4.9.

UCAR Comet, Boulder, Colorado. Milankovitch cycles, Fig. 10.2.

Wiley and Sons, *J. Climatol.*, 1984, Whittaker and Horn: 4, p. 300, Fig. 2a and p. 306, Fig. 4a

Wikipedia

World Meteorological Organization, Geneva

K. Wyrtki, 1985, *The Global Climate System: A Critical Review of the Climate System During 1982–1984*, WMO, Geneva, p. 15, figure 11.

Introduction

Automatic weather station in the Apennine Hills near Madigliana, Romagna, Italy.

WE BEGIN by defining climate and weather and then explaining their significance in everyday life. This helps clarify the scope and importance of climatology and its place in environmental sciences. Following a brief introduction to climatic data and frequently used statistics, we review the history of world climatology over the last two centuries. Finally, we examine the contribution of weather satellites to the study of weather and climate beginning in the 1960s.

1.1 Climate and weather

The word *climate* is derived from the Greek word "klima," meaning slope, and was linked to temperature gradients from Equator to Pole. It entered the English language from French in the thirteenth century. Its modern meaning evolved in the sixteenth century.

Climate is the sum total (or composite) of the weather conditions that generally prevail at a place or over a region. It encompasses the statistics (means, variability, and extremes) of temperature, humidity, atmospheric pressure, wind velocity, cloud cover, precipitation, and other meteorological variables over a long period of time. In contrast, weather is the condition of these same elements and their variations over time intervals of a few days. Conventionally, weather extends out to about 10–15 days – the limit of numerical weather prediction – while longer intervals, typically a month, are considered as part of climate. The "standard" interval used to define climatic characteristics by the World Meteorological Organization (WMO) is 30 years, and these data are called "normals." This term was first used for 1901–1930; the current normal is 1961–1990 or 1981–2010, depending on data availability. However, world weather records were earlier published for 1881–1920. The 30 years must be consecutive and the averages are unweighted. The normals are updated by national climate organizations and the WMO each decade.

Another, much broader definition, that recognizes the complexity of climate, was proposed by the US Committee for the Global Atmospheric Research Program in 1975. It refers to the climatic state as "the average (together with the variability and other statistics) of the complete set of atmospheric, hydrospheric, and cryospheric variables over a specified period of time (monthly, seasonal, annual, decadal) in a specified domain of the earth–atmosphere system." Hydrospheric variables refer to all components of the global water cycle, while cryospheric variables refer to all forms of snow and ice. An updated version of this would also include the biospheric variables on land and in the ocean that affect transfers of energy, water, momentum, and gases between the surface and the atmosphere. The complexity of the climate system is illustrated in Figure 1.1, showing the atmosphere, oceans, hydrosphere, biosphere, and cryosphere. These components, their changes, and their interactions are the subject of this book. Hence, climate is a key element of the global and local environments that has enormous influence on most aspects of our daily lives, whether we live in rural or urban areas. It determines what crops can be grown, how much water is available for drinking and irrigation, and what kinds of shelter and clothing we need.

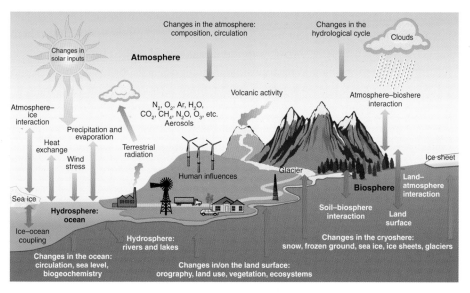

Figure 1.1 The climate system (from Solomon *et al.* 2007, p. 767, Fig. 10.9).

1.2 Why climate matters

Weather affects our day-to-day life and decisions – whether we need an overcoat, umbrella, snow boots, or sunscreen and, in the case of severe weather like thunderstorms or tornadoes, when and where to take shelter. Weather also has major effects on all forms of transportation on land, at sea, and in the air. Climate influences what kind of house we build or buy, whether it has air conditioning or double-glazed windows, and whether it has a flat or sloping roof. It also determines the nature of the local vegetation and agriculture, whether irrigation is needed or not, and the character of the water supply.

Until the mid twentieth century it was considered that climate was essentially constant, but in the 1950s it became widely recognized that as well as ice age events in the distant past, there were important fluctuations on decadal to centennial time scales. We examine changes on these various time scales in Chapter 10. It is also now known that human activities are largely responsible for global warming that began in the late nineteenth or early twentieth centuries and has accelerated over the last few decades. There has been an increasing incidence of heat waves and droughts, as well as floods in other areas. The use of refrigerants (chlorofluorocarbons, or CFCs) led to the formation of the ozone hole in the Antarctic stratosphere, above 15 km, that allows dangerous levels of ultraviolet radiation from the sun to reach the surface in southern South America and Antarctica. While the Montreal Protocol, adopted by most nations of the world in 1987, regulated substances thought to destroy stratospheric ozone, there is still no similar international agreement to regulate gases and pollutants in the atmosphere that cause global warming. Apart from the immediate climatic effects, warming is causing glaciers to shrink and disappear, with consequences for water supplies and global sea level; Arctic sea ice is retreating and thinning; the Greenland and Antarctic ice sheets are losing mass; and plant and animal species, including pests and diseases, are shifting to new areas. For all these reasons, knowledge and under-standing of climate and climatic processes are key aspects of environmental science.

1.3 Climate statistics

The standard statistics that are used in climatology are the arithmetic mean (or average) and a measure of the variability such as the standard deviation (see Box 1A.1). For example, temperature values that vary widely from day to day have

Box 1A.1 Climatological statistics

We provide here a brief summary of the principal statistical measures that are used in climatology. The arithmetic mean ($\bar{x}$) is the sum of the values, x_i, divided by the number of observations (N):

$$\bar{x}_i = \sum x_i/N.$$

The variability of a record is usually specified by the standard deviation, which is the square root of the average sum of squares of the departures of the individual values from the mean (S_N):

$$S_N = \sqrt{\frac{1}{N}\sum_{i=1}^{N}(x_i - \bar{x}_i)^2}.$$

One simple measure of variation is the range – the difference between the maximum and minimum values, either averages or absolute extremes; thus there is a daily range, a monthly range, and an annual range. Another measure is the interquartile range – the difference between the 75th and 25th percentiles in a ranked sequence of the values.

The coefficient of variation (CV) is given by the standard deviation divided by the mean, sometimes expressed as a percentage. This provides a relative measure of variability. It is only suitable for ratio data, such as rainfall amounts (not interval data such as temperature), and it is unreliable if the mean is close to zero, or the distribution of values is not roughly normal. Values of the CV of annual rainfall in western Europe are typically 10–20 percent.

a high standard deviation, whereas those that vary little have a low standard deviation. These values are acceptable for variables such as radiation, temperature, pressure, and moisture content, which are more or less normally distributed (i.e., they cluster around the mean value), but are unsuitable for elements like precipitation and wind speed, which are bounded by a lower value of zero, or cloud amounts that have lower and upper bounds (clear skies and overcast). In such cases the median (central 50 percent value of a numerically ranked distribution) and quartile (25 and 75 percent) values may be appropriate measures of the average value and the range of variation. A further measure of average is the most frequent value – the mode. This is obtained by grouping frequency values into classes that are separated by equal intervals (e.g., 0–5, 6–10, 11–15, etc.) and determining by inspection the class with the highest frequency.

To highlight the point made above, an illustration of the characteristic frequency distributions of common meteorological variables is shown in Figure 1.2. A frequency distribution is a plot of the absolute or percentage frequencies of a data set on the vertical axis against equal-sized groups of the values on the horizontal axis. It should be noted that frequency distributions change shape as the averaging period changes. Thus, for example, hourly precipitation amounts are highly negatively skewed (toward zero), whereas annual totals are closer to a bell-shaped normal (also called Gaussian) distribution. In a normal distribution 95.4 percent of values lie within ± 2 standard deviations of the mean and 99.7 percent lie within ± 3 standard deviations. To isolate outliers, which may be erroneous, in a data set a threshold of four standard deviations is sometimes adopted to screen the data for closer examination. In a normal distribution the mean, median, and mode are equal.

The data that go into climatic "normals" may be hourly values for daily averages, daily values for monthly averages, and monthly values for annual averages. In the case of daily averages, these may be based on 24 hourly values, four six-hourly values, or the average of the daily maximum and minimum readings. Each of these will give a different mean daily value. It should be noted that the basis for such determinations has often changed over time and so a careful check of the station information (or metadata) describing such practices is needed before processing averages for long-term data sets. Missing data can be treated by substituting long-term average values, which does not bias the mean, or by various interpolation or extrapolation approaches using, respectively, data from the station in question, or data from surrounding stations.

A further consideration is the spatial averaging of climatological data. Station data are irregularly distributed in space, whereas model output and satellite data are usually gridded. Station data need to be interpolated to produce spatial maps; this can be accomplished manually by a skilled analyst or by employing a numerical smoothing routine. Such interpolation and mapping is relatively

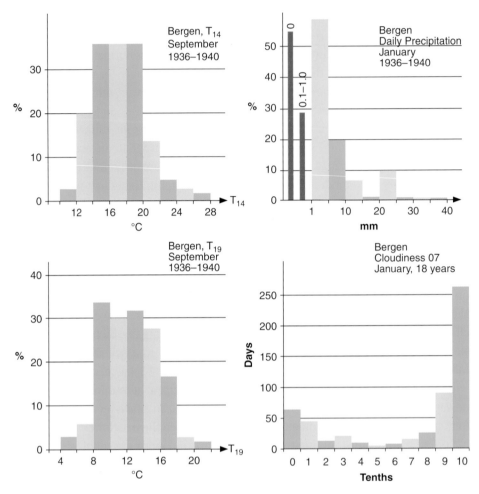

Figure 1.2 Typical frequency distributions of selected meteorological variables at Bergen, Norway (air temperature at 14:00 and 19:00 hours, daily precipitation totals, and cloud amount) (source: after Godske 1966).

straightforward in lowland or plateau areas, but much more difficult in complex and mountainous terrain where the vertical gradients of climatic elements vary according to the variable under consideration, as discussed extensively by Barry in *Mountain Weather and Climate* (2008).

1.4 History of world climatology

Scientific studies of meteorology began when instruments were invented to measure weather elements, notably pressure (the barometer) and temperature (the

Box 1B.1 **Earliest meteorological network**

The earliest meteorological station network was organized in Bavaria in 1780 by the Societas Meteorologica Palatina, based at the Mannheim Academy of Sciences. It operated until 1795, collecting records annually and publishing them over 1781–1792. At its greatest extent there were simultaneous records at 31 stations. For at least one year there were 37 stations, stretching from the Urals, through Europe, to eastern North America.

The instruments and data were standardized. The system anticipated the state networks that developed some 60 years later.

thermometer). The mercury barometer was invented by E. Torricelli in 1643. The mercury thermometer with a scale was invented by D. G. Fahrenheit in 1724. These instruments evolved during the seventeenth and eighteenth centuries, respectively, mainly in Europe. The earliest daily barometric pressure readings were made in Pisa, Italy, in 1657–1658. The earliest temperature readings were made in Uppsala, Sweden, beginning in 1722, with calibrated readings (using the Celsius scale) in use from 1739. Rainfall is reported to have been measured in Greece in 500 BC and in India a century later, but the first records, starting from 1725, were assembled in Britain in 1870 by G. J. Symons. A useful overview of meteorological observations and instruments and their characteristics is provided by Strangeways in his book *Measuring the Natural Environment*. The earliest network of weather stations was set up in the late eighteenth century (see Box 1B.1). Standards for instruments and their exposure were agreed internationally in the late nineteenth century when weather data began to be exchanged via the telegraphic networks, which enabled the collection of weather reports to nearly keep pace with the weather.

Knowledge of world climate also evolved from the late seventeenth to nineteenth centuries. In 1686, British scientist Edmond Halley published a map of the trade winds in the tropical oceans, based on a voyage to the Southern Hemisphere. In 1817, the famous German geographer Alexander von Humboldt published the first map of annual mean temperature in the Northern Hemisphere showing isotherms (lines of equal temperature). In 1832 it was debated at the annual meeting of the British Association for the Advancement of Science whether pressure was the same over the globe. However, a decade later latitudinal differences of air pressure were being reported and discussed.

The term climatology, meaning the science of climatic zones or regions, first came into use in the 1850s. Scientific study of world climatology began in the mid-nineteenth century when sufficient meteorological observations became available to construct world maps of selected variables – pressure, winds, and temperature being among the first. In Germany in 1848, Heinrich Dove produced world maps of monthly mean temperature. In Great Britain in 1860, Robert FitzRoy used telegraphic weather reports to produce the first synoptic weather maps. A *synoptic weather map* is a generalized view of the weather conditions over a region at a fixed time. It shows lines of pressure, winds, temperature values, clouds, and weather conditions. The winds of the Northern Hemisphere were analyzed in 1853 by J. H. Coffin, who went on to carry out a global analysis two decades later. In the 1860s, A. Buchan surveyed the mean global pressure distribution and the prevailing winds on a monthly basis. In 1882, E. Loomis prepared the first map of global precipitation and in 1886 Teisserenc de Bort published the first world maps of monthly and annual cloudiness. In 1897–1899, J. Bartholomew published an atlas of meteorology containing world maps of:

- mean annual temperature and monthly temperatures;
- anomalies, extremes, and ranges of temperature;
- isobars (lines) of mean annual pressure, and mean monthly pressure and winds;
- monthly isonephs (lines of cloudiness);
- monthly isohyets (lines of rainfall);
- monthly storm tracks and frequency;
- annual frequency of thunderstorms;
- average typhoon tracks.

In addition, the atlas contained numerous continental and regional charts.

A further step was taken when *climatic classifications* were developed, notably by Vladimir Köppen around 1900 (see Box 1B.2). His first complete version was published in 1918. The classification combines thermal and moisture categories in a globally applicable scheme that takes account of seasonal characteristics. The boundaries between categories were obtained by determining temperature and moisture values that closely corresponded with vegetation boundaries, such as the Arctic tree line with a July mean temperature of 10 °C. A new version of the Köppen world map is presented in Figure 1.3. The authors also provide maps for each continent.

The five major categories of climate are:

A equatorial climates
B arid climates
C warm temperate climates
D snow climates
E polar climates.

Box 1B.2 Vladimir Köppen

Köppen was born of German descent in St. Petersburg, Russia, in 1846. After studying in Russia he spent most of his life in Germany and Austria. In 1875–1879 he was the chief of the new Division of Marine Meteorology at the German naval observatory (Deutsche Seewarte) based in Hamburg. There, he was responsible for establishing a weather forecasting service for the northwestern part of Germany and the adjacent sea areas. He then moved to a career in research and began working on the relationships between climate factors and vegetation types in the 1880s. He published a first version of his classification of world climate in 1900 and the final version in 1936. In 1924 he and his son-in-law Alfred Wegener published a paper on the climates of the geological past, providing crucial support to the newly published Milankovich theory on Ice Ages, although this was not accepted until the 1960s (see Chapter 10). He died in Graz, Austria in 1940.

The five categories are each subdivided according to the precipitation seasonality (s = dry summer, w = dry winter, f = no dry season). In the equatorial category there is also a monsoon subdivision. The arid climates are subdivided according to degree of aridity into desert (BW) and steppe (BS). The polar climates are subdivided based on temperature into tundra (ET) and frost climate (EF). The characteristics of the climatic types are described in Chapter 9.

A major synthesis of *The Climates of the Continents* was prepared by W. C. Kendrew in 1922 in Great Britain, and appeared in a fifth edition in 1961. In Germany, similar works were published by Köppen in 1923 and Köppen and Geiger in 1930. More recently, the *World Survey of Climatology*, in 16 volumes, was edited by Helmut Landsberg. This includes climatic information about all regions of the world and information about the physical processes of the climate system. The first study of microclimates – Rudolf Geiger's *Climate near the Ground* – appeared in German in 1927; a sixth, English, edition was published in 2003. The first text on physical climatology was published by W. Sellers in 1965 with a more recent work by D. Hartmann in 1994. The first textbook on synoptic climatology – the study of climate as described by airflow and circulation patterns – was written by R. G. Barry and A. H. Perry in 1973. Twenty years later, B. Yarnal wrote a primer on synoptic climatology and, in 2001, Barry and A. M. Carleton published *Synoptic and Dynamic Climatology*, dealing with the mechanisms of global climate and atmospheric circulation, and its spatial and temporal variations. *A Climate Modeling Primer* was published by A. Henderson-Sellers and K. McGuffie in 1987, with a third edition in 2005.

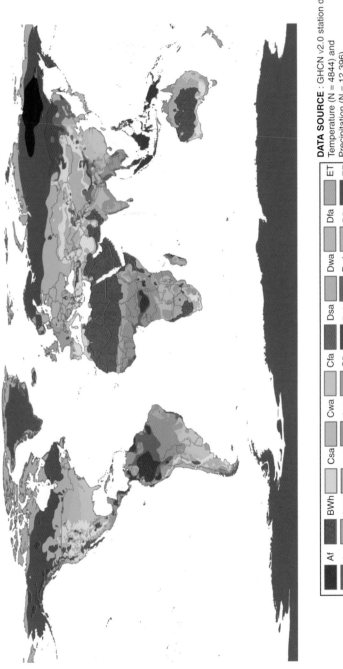

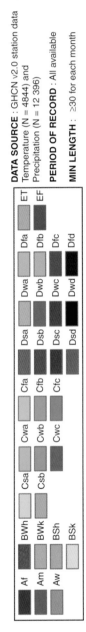

Figure 1.3 World map of the Köppen–Geiger climate classification updated (Peel *et al.* 2007). The map is based on data from the Global Historical Climatology Network (GHCN) that are most representative of the period from 1909 to 1991 for precipitation and 1923 to 1993 for temperature (Source: Peel *et al.* 2007)

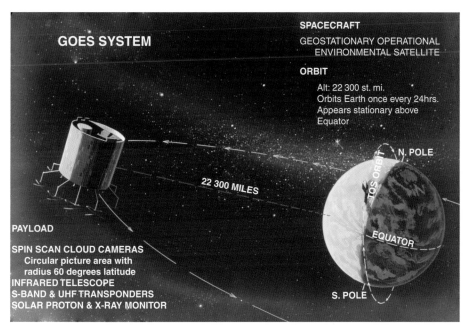

Figure 1.4 The GOES geostationary weather satellite system (source: www.photolib.noaa. gov/bigs/spac0176.jpg).

1.5 The use of weather satellites

The National Aeronautics and Space Administration (NASA) of the United States launched the first *weather satellite* in April 1960 and within a few years global coverage of weather systems in the visible and infrared (thermal) wavelengths became continuous and routine from polar orbiting satellites. Initially, photographic systems were used, and then radiometric sensors, which directly measure the electromagnetic radiation emitted or reflected from clouds or the surface, replaced these. Geostationary satellites were added in 1966, providing repeat views of a given sector of the Earth centered on the Equator every 20 minutes (see Figure 1.4). The US Geostationary Operational Environmental Satellites (GOES) were positioned at 75° W and 135° W from 1974, and in 1977 the Japanese Geostationary Meteorological Satellite (GMS) and European Meteosat were added at 135° E and 0° longitude, respectively. These provide valuable information on the diurnal development of cloud and weather systems. In 1978 atmospheric soundings of vertical temperature and moisture structure became operational, using microwave channels.

Visible channel (0.5–0.8 µm) data are hampered by darkness, which is a major problem in high latitudes during the polar night. Infrared sensors (0.8–12 µm) record temperatures and infrared emission from the surface if there is no cloud or from cloud tops. Passive microwave radiation (1 mm to 3 cm wavelength) emitted by the surface passes through clouds, but is intercepted by heavy rainfall, providing a means to measure it. Radar (*radio detection and ranging*), which is active microwave (3–5 cm wavelength), emits a signal that is bounced back by whatever surface it encounters. It can be used to map sea ice in the polar oceans, as well as areas of rainfall.

There are now ground-receiving stations in more than 170 countries collecting picture transmissions by NOAA satellites. Passive microwave sensors on NASA's Nimbus 5 and 7 in the 1970s–1980s, and on Defense Meteorological Satellite Program (DMSP) satellites from 1979, and multichannel data from the Earth Observing System (EOS) satellites since the 1990s, provide numerous additional global and regional products, including sea ice extent and concentration, cloud cover amount, aerosol concentrations, energy balance components, surface temperature, tropical rainfall amounts, lightning, stratospheric ozone, surface winds over the ocean, and vegetation indices with spatial resolutions of between 1 km and 25 km. Descriptions of available satellite data may be found at:

- http://www.ncdc.noaa.gov/data-access/satellite-data
- http://eospso.gsfc.nasa.gov/
- www.eumetsat.de/
- http://nsidc.org

The first book on satellite climatology, *Climatology from Satellites*, was published by E. Barrett in 1974, and an updated and expanded work was published by Carleton in 1991 under the title *Satellite Remote Sensing in Climatology*. A text on satellite meteorology, *Satellite meteorology: An Introduction*, was published by Kidder and Vonder Haar in 1995.

A separate area of remote sensing is ground-based study of the troposphere. Weather radar has a long history of identifying thunderstorm clouds and areas of heavy rainfall and hail. Techniques evolved after World War II. Now many radar systems are pulsed Doppler that can track the movement of raindrops and precipitation intensity over a sounding range of 150–200 km. Weather radar networks were established in North America, Europe, and Japan between 1980 and 2000.

For research purposes there are also vertically pointing sensors – sodar (sound detection and ranging) and lidar (light detection and ranging). These systems are used to study boundary-layer structure and aerosol layers. Sodar has a vertical range of only 1–2 km, but the systems are inexpensive. Lidar profilers can span the troposphere, but are expensive, like radar.

SUMMARY

We began by differentiating between daily weather and monthly or annual climate, which presents average conditions and their variations at a place, in a region, or globally. The climate is typically calculated for a 30-year period.

The principal climatic statistics are the arithmetic mean and the standard deviation, although for some climatic elements the median and quartile values are preferable.

Climatology developed as a science in the late nineteenth century, establishing subfields in the twentieth century. For regional studies a notable advance involved the classification of climatic conditions based on temperature and precipitation thresholds by V. Köppen. Global views of weather conditions were made possible beginning in the 1960s with the launch of weather satellites, both polar orbiting and geostationary. Remotely sensed data are collected globally in the visible, infrared, and passive and active microwave channels. Vertical profiling of the atmosphere is widely conducted. Numerous global climate products from remote sensing are now available online. Ground-based remote sensing with radar systems is widely used to track severe weather, while lidar and sodar are used in research.

QUESTIONS

1 Compare the everyday applicability of weather and climate information.
2 Compare the use of the mean and the median for characterizing precipitation data.
3 What measures of variability are useful for
   (a) temperature data,
   (b) precipitation data?
4 Discuss the bases of Köppen's classification of climate.
5 How does climate differ from weather?
6 When did climatology begin as a science?
7 Explain the difference between polar orbiting and geostationary satellites and their imagery.
8 Examine and describe the current satellite images at: www.goes.noaa.gov/ www.ospo.noaa.gov/Products/imagery/Index.html

2 The elements of climate
A global view of energy and moisture

Cumulonimbus cloud and a rain shower.

THE ELEMENTS of climate involve external forcing factors – notably solar radiation – response functions such as air temperature, and moisture variables like precipitation and moisture content. It is not possible to follow a totally logical sequence due to the interactions between many of the variables. Nevertheless, we shall begin with solar energy and other forms of energy, then temperature, and proceed from there to moisture variables such as clouds, precipitation, and evaporation.

Weather variables are measured at a global network of about 7000 stations every six hours and these data are collected and analyzed to produce climatic statistics for hourly, daily, monthly, and annual intervals. These and other historical data are available at: www.ncdc.noaa.gov/oa/climate/climatedata.html

Climatic variables such as solar radiation are recorded by much more limited specialized networks.

2.1 Energy

(a) Solar radiation

The energy that warms the Earth and its atmosphere, thereby supporting life, and also drives the atmospheric circulation, is supplied by *solar radiation*. The sun emits electromagnetic energy from its photosphere (the lowest layer of the solar atmosphere, effectively the visible solar surface), which has a temperature of about 6000 °C. Solar energy is emitted at a rate proportional to the fourth power of the absolute temperature, meaning that the sun emits enormous quantities of energy. The short wavelength of this radiation is in the range 0.1–3 μm, or micrometers (μ = 10^{-6}, this Greek letter is pronounced "mew") with a maximum in the visible light band at 0.5 μm (see Box 2A.1). Ultraviolet radiation is shorter than 0.4 μm, and most of this is absorbed by atmospheric ozone (O_3) gas between 10 and 50 km in the upper atmosphere (see Box 2B.1). Visible radiation (light) is emitted between 0.4 μm (blue light) and 0.8 μm (red light) (see Box 2B.2), and infrared (or thermal) radiation between 0.8 and 3 μm. The amount received at the top of the atmosphere (ToA) on a surface at right angles to the beam is ~ 1366 W m^{-2}. (Watt (W) is the international unit of power or work done; W = joules s^{-1}.) This value is known as the "solar constant." Recent data from NASA's Solar Radiation and Climate Experiment (SORCE) indicate a new lower value for total solar irradiance of 1361 W m^{-2} for the solar minimum stage of the ~ 11-year solar cycle.

The solar emission fluctuates over the 9–12-year solar cycle of sunspots by about ± 1 W m^{-2}. Sunspots are dark, cooler areas surrounded by brighter peripheral rings that are hotter. The annual number of sunspots ranges between about 50–160 at maxima and 10–20 at sunspot minima. The interception by the area of the Earth's disc is spread out over the spherical surface of the Earth ($\pi R^2 / 4 \pi R^2$), and this translates into a global average of 342 W m^{-2}, or one-quarter of the solar constant.

The ToA solar radiation is symmetrically distributed meridionally (i.e., with latitude) and the pattern shifts north and south with the seasons. The pattern is symmetrical about the Equator at the vernal (spring) and autumnal equinoxes (when the sun is overhead at noon on the Equator). The seasons are a result of the annual revolution of the Earth about the sun and the fact that the Earth's axis is tilted by approximately 23.4° from the vertical. In boreal summer (June) the North Pole is tilted toward the sun and in austral summer (December) the South Pole is toward the sun. The tilt of the axis leads to the sun being higher in the sky during the respective summer seasons. The sun is overhead at noon on 21 June at the

Box 2A.1 Radiation

The wavelength of maximum radiation (λ_{max}, pronounced lambda) varies inversely with the absolute temperature according to Wien's Law:

$$\lambda_{max} = (2897/\,T)\ 10^{-6}\ m.$$

Hence, solar radiation has a maximum wavelength at 2897 / 6000 = 0.48 μm and terrestrial radiation has a maximum at 2897 / 288 = 10.1 μm.

The sun emits radiation as a "black body" – one that absorbs and emits the maximum possible radiation at all wavelengths. The emission is proportional to the fourth power of the absolute temperature in kelvin (K), T^4. This is known as the Stefan–Boltzmann relationship. The value of 0 K corresponds to –273 °C; the freezing point of water is 273 K or 0 °C, and the interval of 1 K is equal to 1 °C. The kelvin scale does not use a degree symbol.

The Earth is not a black body and each surface has an infrared emissivity (ε, epsilon) value of less than one. Values of ε range from 0.92 to 0.98 for water, 0.90 for desert, 0.90 for fields and forest, and 0.82–0.99 for fresh snow.

Box 2B.1 Atmospheric structure

The atmosphere forms a very thin skin around the Earth and is about 15 thousandths of the Earth's radius. Of this tiny fraction, humans can inhabit only the lowest 5 percent (the lowest 5 km).

The atmosphere is conventionally divided into a number of vertical layers. The lowest layer where most weather phenomena occur is the *troposphere* (turbulent sphere), which is about 8 km deep at the Poles and 15 km deep in the Tropics due to deep cloud convection. Temperature and moisture generally decrease with altitude in the troposphere. At the top of the troposphere is a layer known as the tropopause, where temperatures cease decreasing (see Figure 2.1). This came as a total surprise when the first balloon ascents were made around 1900.

The second layer is the *stratosphere* (stable sphere) where temperatures begin to increase with height as a result of the absorption of ultraviolet radiation by atmospheric ozone at around 20–40 km altitude. There is very little moisture in the stratosphere. At about 50 km altitude we encounter the base of the mesosphere (middle sphere) where temperatures again decrease with height. Finally, at about 80 km, temperatures again begin to increase with height in the thermosphere (heat sphere) as a result of the absorption of extreme ultraviolet radiation by molecular and atomic oxygen. However, in this rarified atmosphere temperatures are largely theoretical. The ionosphere (ionized sphere), where aurorae develop in the two polar regions, overlaps mainly with the thermosphere, but extends out to 600 km above the surface into the exosphere (outer sphere), where hydrogen atoms are very sparse.

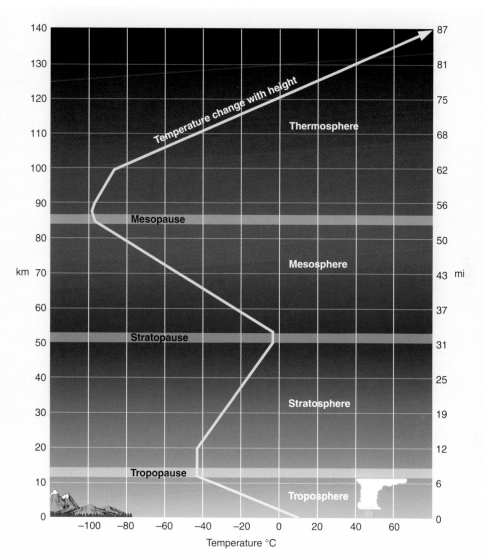

Figure 2.1 Schematic diagram of the temperature structure of the atmosphere, showing the layers into which it is divided. Heights are in kilometers (left) and miles (right) (source: NOAA).

Tropic of Cancer (23° N latitude) and on 22 December at the Tropic of Capricorn (23° S latitude). At the Equator the sun is overhead at noon on 20 March and 23 September (the equinoxes). The winter and summers seasons in terms of heat and cold lag about a month behind the sun's passage due to the time it takes for the surface to heat up and cool down (see Chapter 8). While middle latitudes experience four seasons, defined by temperature, seasons in the Tropics are determined by moisture conditions (rainy and dry seasons).

Box 2B.2 Why is the sky blue?

Light spans the visible spectrum from 0.4 μm (blue light) to almost 0.8 μm (red light). The shorter blue wavelengths are scattered by air molecules about four times more than the longer red wavelengths, so the sky normally appears blue when seen from the ground. This molecular scattering (or Rayleigh scattering) is almost equal in all directions. When there are water droplets (fog or cloud) present, the scattering is almost all directed forward towards the ground; this is known as Mie scattering.

As we go higher in the atmosphere, the air density decreases and there are fewer molecules of air present. Because there are fewer scatterers, the sky becomes black.

In addition to the effects of latitude and season on solar radiation receipts, cloud cover that reflects the incoming radiation and atmospheric aerosols that reflect and absorb it help to determine the surface distribution of incoming solar radiation. These factors are discussed below.

Solar radiation at the top of the atmosphere is measured by satellite spectro-radiometer (see Glossary). At the surface, pyranometers are used that comprise a blackened thermopile covered by a glass dome that excludes infrared radiation. A thermopile is an electronic device that converts thermal energy into electrical energy. It is composed of several thermocouples, usually connected in series. The typical measurement range of a pyranometer is 0.3–2.8 μm.

A further climatological measurement is sunshine duration (in hours). This is determined with a sunshine recorder (known as a Campbell–Stokes recorder), where a glass ball focuses the sun's beam to burn a trace on a strip chart that is changed daily. The instrument was invented in 1853. The burning on the chart corresponds to a threshold value of solar radiation averaging 120 W m^{-2}. Increasingly, automated recording systems are being employed.

(b) Cloud effects

Clouds mainly reflect solar radiation, but there is also a small component that is absorbed. Cloud reflection (or its *albedo*) is high – in the range 30–80 percent, according to cloud optical depth and its liquid water content. This is why clouds appear very bright from above. The optical depth is a measure of the radiation absorbed or scattered along a path through the cloud, i.e., the cloud transparency. Table 2.1 illustrates typical albedo values for clouds and other surfaces. On average, clouds cover about 66 percent of the Earth's surface, and so, with their high albedo, are a potent factor in reducing incoming solar radiation. Cloud water droplets also absorb a few percent of incoming solar radiation.

Table 2.1 Typical fractional albedo values for natural surfaces

Surface type	Albedo
Ocean	0.06–0.10
Forest	0.10–0.15
Grass	0.25
Desert sand	0.35
Fresh snow cover	0.80–0.90
Global surface	0.14–0.16
Cirrus	0.40–0.50
Stratocumulus	0.60
Cumulonimbus	0.90
Planet Earth	0.30

(c) Aerosol effects

Aerosols are airborne particles and/or liquid droplets and gases together. Fine airborne particles have a typical diameter in the range 0.01–100 µm. Marine and continental background aerosols are in the range 0.1–1.0 µm. Aerosols originate from fires in forests and homes, industrial activity, biogenic matter (from vegetation), sea spray, dust from dry soils, and volcanic eruptions. Some start as particles while others form from gas-to-particle conversion. They include sulfates, nitrates, black carbon, and mineral dust. Natural sources produce about 3600×10^9 kg yr^{-1}, while anthropogenic (human) sources yield $300–900 \times 10^9$ kg yr^{-1}, or some 10–25 percent of natural sources.

Aerosols exert direct radiative forcing by scattering solar radiation both forward towards the surface, and backward to space; they also absorb a small proportion of it. The relative effects are complicated and depend on the physical and chemical characteristics of the aerosols. Aerosols also have microphysical effects on clouds that can increase or decrease cloud amounts depending on the aerosol load and the time of day. From a radiative point of view, the most significant aerosols are black carbon (soot) from forest fires and biomass burning, and mineral dust blown from desert surfaces and dry soils. Particle concentrations in clean continental air are ~ 3000 cm^{-3}, compared with $\sim 50\,000$ cm^{-3} in polluted continental air, of which about two-thirds are soot. Desert air averages ~ 2500 cm^{-3}, and clean marine air 1000–1500 cm^{-3}. Typically, concentrations decrease exponentially with altitude, but volcanic eruptions lift fine ash and sulfates into the stratosphere. Here they circle the globe, eventually settling out over a 2–3-year period. Long-distance ash transport was dramatically illustrated in April 2010, when ash from a

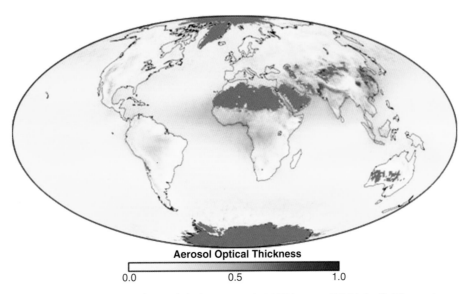

Figure 2.2 Average aerosol optical thickness, 2000–2007 (source: NASA Earth Observatory; http://earthobservatory.nasa.gov/IOTD/view.php?id=8857).

volcanic eruption in Iceland grounded aircraft in Europe for six days and another eruption in May 2011 had a similar but more brief effect.

The *aerosol optical depth* (or thickness) is defined as the integrated extinction coefficient over a vertical column of unit cross-section; extinction coefficient is the fractional depletion of radiance per unit path length. The absolute range is 0.0–1.0. Optical depth ranges from 0.5 to 1 over eastern Asia, the Middle East, and North Africa (Figure 2.2).

On average, about 20–25 percent of the optical depth is contributed each by sulfate, dust, sea salt, and particulate organic matter, with a few percent by black carbon (soot).

Optical depth has been measured since 1993 at a global land network (Aeronet) of about 200 stations, and since 1979 by satellite sensors in the visible and near-infrared bands. There have also been a number of field campaigns in various parts of the world with ground and airborne measurements.

(d) Surface receipt of solar radiation

The amount of solar radiation reaching the Earth's surface depends on the latitude, season, time of day, cloudiness, aerosol backscatter, and surface geometry. The absorbed short-wave radiation (Figure 2.3) has maximum values over 300 W m^{-2}

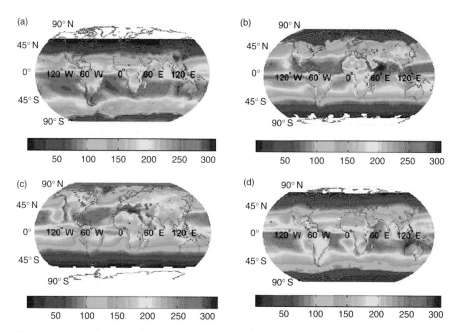

Figure 2.3 Net downward solar radiation (W m^{-2}) absorbed at the surface for (a) January, (b) April, (c) July, and (d) October, 1984–1997. The values range up to 300 W m^{-2} (source: Hatzianastassiou *et al.* 2005).

at the Equator and minimum ones of 80 W m^{-2} at the Poles. The global average is approximately 150 W m^{-2}, corresponding to 44 percent of that incoming at the top of the atmosphere.

The amount absorbed is a function of the surface reflectivity, or albedo. This ranges from about 0.06 for the ocean, 0.10 for forest, 0.20 for agricultural land, 0.35 for desert sand, and up to 0.9 for fresh snow cover (Table 2.1). Hence, the global patterns of absorbed solar radiation are seasonally variable and geographically complex.

(e) Infrared radiation

The Earth's surface is heated to ~ 20 °C on average by the absorption of radiation, and as a consequence it emits thermal *infrared radiation* (IR) with wavelengths of 5–50 μm and a maximum at about 10 μm. In the wavelength range 8–12 μm there is an "atmospheric window" where the atmosphere is largely transparent, allowing infrared radiation to escape to space. On either side of this window there are absorption bands due to atmospheric gases, notably water vapor and carbon dioxide – both so-called *greenhouse gases* (see Box 2A.2).

Box 2A.2 Atmospheric chemistry

The atmosphere is a mixture of gases, notably nitrogen, N_2 (78 percent) and oxygen, O_2 (21 percent) for dry air. However, these play little role in weather and climate and there are many important trace gases that do – carbon dioxide (0.039 percent), ozone (0.00006 percent), and methane (0.00017 percent). The last three are greenhouse gases, meaning they admit solar radiation but absorb outgoing infrared radiation, leading to a heating of the atmosphere. The greenhouse analogy is inexact because while the glass in a greenhouse admits solar radiation, the heating effect is a result of the trapping of sensible heat (warm air) by the glass, not the trapping of infrared radiation as in the atmosphere. The primary greenhouse gas is water vapor, which averages about 1 percent by volume in a moist atmosphere, but is highly variable. It accounts for 50 percent of the natural greenhouse effect, together with water drops in clouds (25 percent), and carbon dioxide (CO_2) (20 percent). The balance is accounted for by methane, nitrous oxide, and chlorofluorocarbons. Note that the "natural" greenhouse raises the mean atmospheric temperature by about 30 °C above that of a dry atmosphere, and the human-induced greenhouse, due to burning fossil fuels, accounts for current global warming of about 1 °C.

Ozone (O_3) is created in the upper stratosphere (40–50 km altitude) by the action of extreme ultraviolet radiation (0.1–0.2 μm) in breaking down molecular oxygen (O_2) into atomic oxygen (O) and the O_2 and O combining to form ozone. Ozone is destroyed in the lower stratosphere (15–20 km) by interaction with chlorofluorocarbons (CFCs), which are now banned. However, CFCs have lifetimes of 45 and 110 years, meaning they will remain in the stratosphere for years to come. The ozone hole that forms each spring (September–October) in the Antarctic stratosphere develops as ozone is destroyed by chemical reactions with chlorine and nitrous oxides. It was discovered at Halley Bay, Antarctica, by British scientists in 1985, but originated in the 1970s. It is currently about 27 million square kilometers in September–October of each year and will not recover to its former state until around 2050. In 2011 a large ozone hole was observed in the Arctic for the first time as a result of low temperatures in the stratosphere. Ozone is also present in small amounts at ground level as a result of the action of sunlight on hydrocarbons and nitrogen oxides, which are emitted by automobiles, fossil-fuel power plants, and refineries.

CO_2 is formed naturally by the respiration of biota and soil microbes and is released from the Earth's interior. Its anthropogenic (human) production is from fossil fuel burning, forest clearance and burning, and cement manufacture. It is dissolved in the ocean and consumed in plant photosynthesis, but there is a net gain by the atmosphere. This currently amounts to almost two parts per million (ppm) annually, so the global atmospheric concentration has risen by almost 40 percent from 280 ppm before 1800 to 390 ppm in December 2011. The hydrocarbon *methane (CH₄)* is released by anaerobic processes in wetlands and rice paddies, by enteric fermentation by cattle and termites, and by biomass burning, oil and coal extraction, and from landfills. Methane concentrations rose from 700 ppb (parts per billion) before 1800, to 1825 ppb today. Over a 100-year time scale, methane has a global warming potential that is 25 times that of an equivalent mass of carbon dioxide.

Recently it has been suggested that reducing the release of methane and black carbon could delay global warming and is potentially more readily achievable in the near term than regulating CO_2 emissions.

The atmosphere absorbs much of the outgoing IR radiation and re-radiates it back to the surface and to space. This process happens between individual atmospheric layers. On average, the cooling rate due to infrared radiation losses from the atmosphere between 1 and 8 km altitude is about $2\,°C\,day^{-1}$. This is offset by absorbed solar radiation and heat transferred from the surface. Under clear skies nearly all of the atmospheric back radiation to the surface originates in the lowest 1000 m of the atmosphere. It is common to consider the net IR radiation (at the surface or at the top of the atmosphere) that is almost invariably negative. The global pattern of net IR is generally smooth and zonally orientated (east–west). There is a maximum of $-280\,W\,m^{-2}$ over the Sahara, where surface temperatures are highest due to the clear skies and desert surface, and minima of $-180\,W\,m^{-2}$ over the Poles, where it is coldest.

Infrared radiation is measured by a pyrgeometer, which consists of a thermopile covered by a silicon dome coated to exclude solar radiation and transmit radiation between about 4 and 50 μm.

(f) Net radiation

Net radiation (Rn) is the sum total of shortwave and IR radiation. It can be written:

$$Rn = S\,(1 - \alpha) + (I{\downarrow} - I{\uparrow}),$$

where S = incoming solar radiation, α = albedo, I↓ and I↑ are the downward and upward IR fluxes, respectively. On average, the net radiation over most land surfaces ranges between about 0.5 and 0.65 of the incoming solar radiation amount. At night the net radiation is usually negative because there is only net outgoing infrared radiation.

Net radiation at the surface has a relatively straightforward distribution pattern over the globe, as shown in Figure 2.4 for (a) July and (b) December 2006. Monthly amounts range from $\sim 250\,W\,m^{-2}$ in the respective Tropics to near zero over the winter Pole.

Net radiation at the surface is measured with a net pyrradiometer that consists of upward- and downward-facing thermopiles that measure the difference between the downward and upward fluxes of all-wavelength radiation. Semi-rigid poly-thene domes protect the sensors.

(g) Heat fluxes

Heat flux is the rate of energy transfer through a given surface. It is measured in Watts (W), or per unit area in $W\,m^{-2}$ (see Appendix on units). Fluxes can be

Figure 2.4
Annual net
radiation over the
globe at the
surface in (a) July
2006 and (b)
December 2006
from the Cloud
and Earth's
Radiant Energy
System (CERES)
(source: NASA
Earth
Observatory).

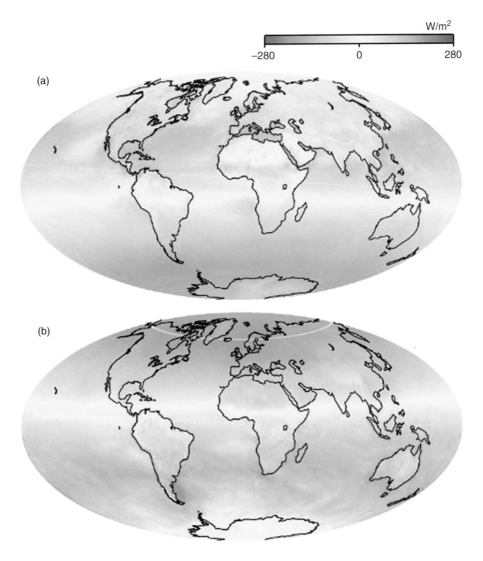

radiative, conductive, or convective. Here we are concerned with the conductive and convective fluxes of heat from the ground surface. The surface is an important source of heat for the atmosphere. First, *sensible heat* (H) is transferred by conduction to an air layer a few millimeters thick. Conduction involves microscopic diffusion of heat as the result of a temperature gradient. Then, the heat is transferred to the boundary layer (~ 100–1000 m deep) by warm air rising in convective plumes, or by being mixed upwards in wind gusts. Convection involves bulk motion. Sensible heat is determined by the vertical temperature gradient near the surface and the horizontal wind speed.

Second, *latent heat* (LE) is transferred upward following the evaporation of moisture at the surface. This heat is released when condensation occurs at the cloud condensation level in the amount of 334 kJ kg^{-1}. Latent heat is determined by the vertical moisture gradient near the surface and the horizontal wind speed. Alternatively, it can be determined from the energy required for water to evaporate from the surface. Obviously, LE is greatest over the oceans due to the unlimited water supply, while H is greatest over heated desert surfaces. Mean values of LE reach 200 W m^{-2} off the east coast of the United States, where cold air in winter flows over the warm Gulf Stream, and likewise off the east coast of Asia, while H reaches 80 W m^{-2} in the Sahara (Figure 2.5).

There is also heat flux into or out of the ground by conduction (G). Conduction in the soil involves heat transfer within or between bodies of matter. The magnitude of this in the ground depends on the temperature gradient in the soil and the thermal properties of the soil. It is generally small.

The surface *energy balance* can be written:

$$Rn = H + LE + G.$$

Rn is positive toward the surface; the heat fluxes H, LE, and G are here defined as positive away from the surface.

For land surfaces, H is typically about 40 percent, LE about 50 percent, and G about 10 percent of the net radiation. However, there is substantial seasonal variation. For a deciduous forest in Russia, the latent heat changes from 65 percent in summer to 10 percent in winter due to the loss of leaves cutting off transpiration. Sensible heat in forests is an energy source in autumn–winter and a sink in spring–summer.

There is also diurnal variation. Typically, H and LE follow the daily curve of net radiation with peaks near midday. They are directed toward the surface at night to offset the negative net radiation due to outgoing infrared radiation. The small ground heat flux term more or less balances out over the daily cycle as there is little heat storage in the ground. Over the annual cycle, heat flows into the ground in summer and out in winter (although it may be trapped in the ground by deep snow cover).

An important index of the relative roles of sensible and latent heat is the Bowen ratio ($\beta = H/LE$). $\beta = 1$ when the sensible and latent heat terms are equal; it is greater than 1 for a dry surface and less than 1 for a moist surface.

(h) Temperature

Air temperature is a measure of the atmospheric heat content. It is a response to the combined effects of absorbed solar radiation by the ground surface, the vertical

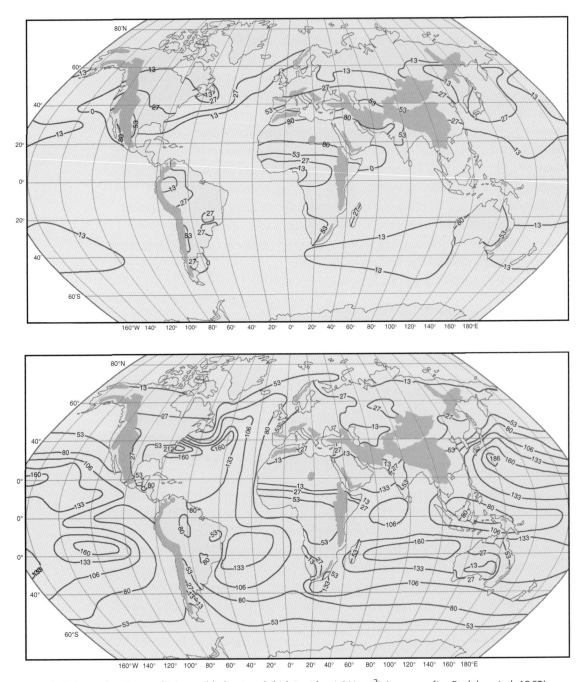

Figure 2.5 Annual patterns of (a) sensible heat and (b) latent heat (W m^{-2}) (source: after Budyko *et al.* 1962).

transfer of sensible and released latent heat to the air by convection, and the horizontal advection (movement) of warm and cold air masses. The standard measure is the air temperature at a height of 2 m above the surface, typically measured by a mercury thermometer in a Stevenson screen or weather shelter – a white-painted, louvered box on a stand. Temperature change on a daily to weekly scale is recorded by a thermograph, which measures the temperature via a bimetallic strip that is connected by a series of levers to a pen that traces the temperature on a chart on a rotating drum. The chart may be daily or weekly.

Mean monthly temperatures over the globe show a basic zonal (west–east) pattern interrupted by the continents and warm/cold ocean currents (Figure 2.6). The presence of clear skies and cloudy skies also plays a role. In January, highest average temperatures are 30 °C in the Congo and the Australian desert. The lowest averages are about − 45 °C in northeastern Siberia. In July, the highest average is 35 °C over the Sahara. In the Arctic the mean July temperature is ∼ 0 °C and at the South Pole, due to the high altitude and the austral winter darkness, it is near − 60 °C.

Important characteristics of temperature are the daily (diurnal) and annual variations (see Section 8.2). The annual cycle of temperatures lags about a month behind the peak solar radiation over land (i.e., the warmest month is July) while sea surface temperatures peak in August or early September due to the higher heat capacity of the ocean and vertical mixing in the water. (The heat capacity is the amount of energy required to raise the temperature of a body by 1 °C.) Correspondingly, the maximum daytime temperatures over land typically occur around two hours after noon, when solar radiation is at a maximum, because the absorbed heat at the surface has to be transferred to the atmosphere by sensible heat. Under cloudless skies, the daily minimum temperature occurs around sunrise since the radiative cooling of the ground is a cumulative process. The diurnal temperature range is greater inland than at the coast and reaches maximum values in desert areas of around 40 °C, compared with only 5–8 °C on mid-latitude coasts.

For many applied purposes, such as agriculture, an accumulated degree-day index can be used. The temperature departure from a threshold value is summed on a daily basis. For example, thawing and freezing degree days are determined with respect to positive and negative departures, respectively, from 0 °C. Thus, a daily average temperature of 5 °C represents five positive degree days, and so on. Growing degree-days in mid-latitudes are often determined with respect to a threshold of 6 °C, and heating/cooling degree-days for building energy use are calculated with respect to a threshold of 18 °C.

The vertical decrease of temperature (the lapse rate) with height averages about 6.5 °C per kilometer in the troposphere (the lowest atmospheric layer). The troposphere extends up to 8 km at the Poles and 15 km in the Tropics, where there is a

Figure 2.6
Mean monthly surface
air temperatures (°C)
for January, April, July,
and October
1961–1990
(source: Jones *et al.*
1999).

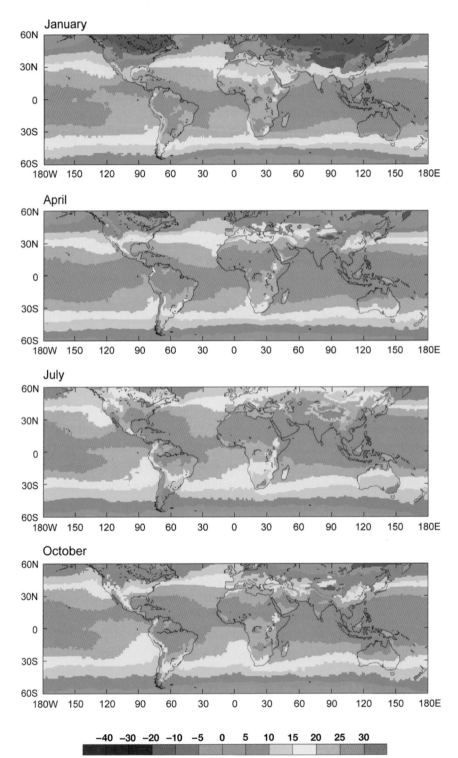

temperature rise at the base of the stratosphere (the tropopause) as a result of the absorption of incoming ultraviolet radiation by the ozone layer. The actual rate of temperature decrease with height is known as the *environmental lapse rate*.

Sometimes, the temperature is lower at the ground surface than in the air above due to strong nighttime cooling by outgoing radiation, from a snow surface, for example. In this case there is an inversion of the temperature change with height and the temperature may increase for hundreds of meters. This pattern is dominant in polar regions, and in mid-latitudes is common at night in winter. Further ways in which temperature inversions may form are, first, when warm air flows over a colder surface layer, and second, when upper-level air subsides in a high pressure system causing warming of the sinking air through compression. Inversions have important effects in capping upward convection and trapping pollutants near the surface.

The *dew point* temperature is the temperature at which air becomes saturated when cooled at constant pressure. It is determined from readings of a dry-bulb thermometer and a wet-bulb thermometer that has a wet muslin bag around the thermometer bulb, so that there is cooling by evaporation. In a psychrometer, air is forced past the two thermometers at a constant speed by a fan. The wet bulb has a lower reading than the dry bulb if the air is unsaturated, and this wet-bulb depression is used with the dry-bulb reading to calculate the dew point and the relative humidity (see Section 2.2a).

A heat index that combines air temperature and relative humidity is often calculated in hot weather in an attempt to determine the human-perceived equivalent temperature – how hot it actually feels. The details of its calculation are available at http://en.wikipedia.org/wiki/Heat_index. Correspondingly, in winter a wind chill index combines the effects of wind speed and air temperature on the skin. Wind chill charts in imperial and metric units are given at http://en.wikipedia.org/wiki/Wind_chill. (See also Section 4.6.)

The definition of a *heat wave* recommended by the World Meteorological Organization is when the daily maximum temperature exceeds the average maximum temperature by 5 °C for more than five consecutive days (the average being the normal period 1961–1990). We shall discuss heat waves in Section 12.1.

2.2 Moisture

(a) Vapor content, vapor pressure, relative humidity, and precipitable water

There are a number of different measures of atmospheric moisture. The atmospheric *vapor content* describes the weight of moisture (g) held in a kilogram of air (g kg^{-1}).

Table 2.2 Relative humidity (percent) from values of dry-bulb temperature and wet-bulb depression

Dry bulb °C	Wet bulb depression °C					
	0.5	1.0	2.0	3.0	5.0	10.0
−10	89	69	39	10	–	–
0	91	82	60	41	15	–
10	94	88	76	55	38	–
20	95	91	82	74	58	24
30	96	93	86	79	67	39
40	97	94	88	82	72	48

This is known as the *specific humidity* if the moisture content of the air is included and as the humidity mixing ratio (HMR) if the air is dry (the two are almost numerically identical). The HMR has a range from about 1–10 g kg^{-1} at the surface and decreases more or less exponentially with altitude. Almost all atmospheric moisture is contained below 500 mb (6 km). Another measure of atmospheric moisture is *vapor pressure* – the partial pressure exerted by water vapor as a gas, which is typically 1–10 mb, so it is a very small fraction of atmospheric pressure (see Section 3.1). The vapor pressure increases non-linearly with temperature (the Clausius–Clapeyron relationship) and has a saturation value that is determined by the air temperature, reaching a maximum value at the boiling point.

The ratio of the vapor pressure to its saturation value at that temperature is termed the relative humidity (0–100 percent). This can be calculated from the dry-bulb temperature and the wet-bulb depression (see Section 2.1h). Examples of this calculation can be found in Table 2.2. Low relative humidity is associated with moderate wet-bulb depressions and low temperatures.

Relative humidity is a less useful indicator of moisture content since its value is affected by air temperature. The relative humidity decreases by day as the air temperature increases, and increases at night when it is cooler (assuming there is no change in the air mass). The daily or weekly change of relative humidity is recorded by a hair hygrograph that was invented by Horace Benedict de Saussure in Geneva in 1783. It consists of a bundle of human hair that is connected via a series of levers to a pen that traces on a chart affixed to a revolving drum. The instrument is not very accurate – within ~ 10–20 percent. Perhaps more useful in climatology is the total column moisture content or *precipitable water content*. This is the depth of the column of water that would be obtained if the water vapor in the atmosphere were condensed out. This value is related to the possible precipitation

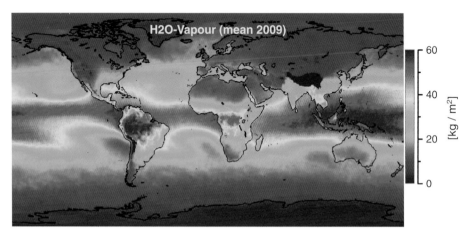

Figure 2.7 Precipitable water content for 2009 (kg m^{-3} = mm) derived from Global Ozone Monitoring Experiment (GOME)-2 from ESA (source: www.globvapour.info).

amount. The global distribution, shown in Figure 2.7, has averages values of about 50 mm near the Equator and < 5 mm in polar regions. The column precipitable water increases by about 7 percent per 1 °C of warming, so this can be a potent factor as global temperatures rise (see Section 10.4).

The horizontal transport of moisture in the atmosphere depends on the vertical distribution of water vapor and the horizontal wind velocity. It is also an outcome of storm tracks. Accordingly, it will be discussed below in Section 3.1c.

(b) Clouds

*Cloud*s comprise suspended water droplets and/or ice crystals. They exist in a wide variety of forms from sheets, to rolls, to vertical towers, and they occur from the surface (fog) to the upper troposphere (10–15 km). High-level cirriform cloud is composed of ice crystals (cirrus is Latin for hair). Low-level stratiform cloud is mainly water droplets (stratus is Latin for layer), and medium-level cloud (alto-cumulus/altostratus) is mainly mixed-phase (ice crystals and water droplets). Cumulus (Latin for heap) is turreted, with its horizontal base at low levels. The Latin names were proposed by Luke Howard in England in the early 1800s to ensure that the same forms were identified by observers worldwide.

Values of cloud liquid water content (LWC) have a range that depends on cloud type: stratus and cumulus 0.25–0.30 g m^{-3}, stratocumulus 0.45 g m^{-3}, and cumulonimbus 1.0–3.0 g m^{-3} (nimbus is Latin for rain). Cloud LWC typically makes up only a small fraction of the moisture in an atmospheric column.

Cloud droplets form when water vapor condenses on a condensation nucleus that has the property of wettability (i.e., it is hygroscopic). The most common hygroscopic nuclei are sea salts and sulfate and chloride particles. Ice crystals form on freezing nuclei that are typically clay minerals that become active at around $-9\,°C$.

In the absence of freezing nuclei, cloud droplets can be super-cooled (cooled below the freezing point) to $-39\,°C$ before they freeze spontaneously. For this reason, middle-level clouds at $-10\,°C$ to $-30\,°C$ are typically a mixture of super-cooled water droplets and ice crystals.

Clouds form in several main ways: by thermal convection, as a result of daytime surface heating, giving rise to vertical cumulus towers; by uplift of air along the sloping boundary between bodies of contrasting air (also known as slantwise convection) giving rise to extensive stratiform cloud sheets; and by orographic uplift that may trigger convection or lead to upstream cloud bands. Local cloud forms also develop through airflow interactions with topography, as in the case of summit cap clouds and lee-wave lenticular cloud bands that form downwind of the mountain barrier.

The global distribution of cloud cover primarily reflects the occurrence of oceanic storm tracks where cloud amounts exceed 80 percent (Figure 2.8; compare to Figure 3.19). By contrast, amounts in January in the eastern Mediterranean/North Africa are only 10 percent. The global cloudiness averages about 66 percent.

Trained observers determine cloud amount, type, and height at a point above the surface by visual means. Cloud amounts are determined separately for low, medium and high cloud, as well as total cloudiness. Amounts are given in oktas, or eighths. Cloud base height at night can be measured with a ceilometer that projects a light beam onto the cloud base. Visible band satellite images readily give cloud amounts over an area, but cloud height has to be estimated from the infrared temperature of the cloud top; low temperature means high cloud.

(c) Lightning

Lightning is a discharge of atmospheric electricity caused when there is a strong electrical field in a thunderstorm cloud between ice crystals – that are positively charged – near the cloud top, and negatively charged soft hail pellets (graupel) near the cloud base. Collisions between ice crystals and graupel lead to the observed charge separation and the lighter ice crystals are transported by updrafts to the cloud top. Negative charge descends from the cloud base toward the ground in a leader and there is a return stroke from the surface. The rapid expansion of the heated air channel causes the thunder. Many aspects of cloud electrification and

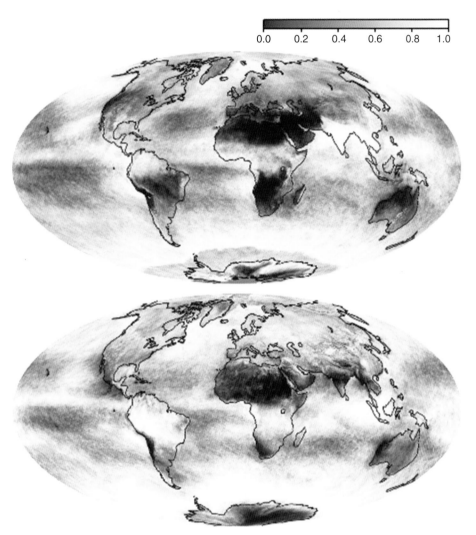

Figure 2.8
The global distribution of cloud cover fraction, 0–1, in (top) July 2006, (bottom) December 2006 (source: NASA Earth Observatory).

lightning remain poorly understood. Lightning also frequently occurs between neighboring clouds.

Peak flash rates are observed over the Congo, Indonesia, and tropical South America (Figure 2.9). During 1998–2000 the North American Lightning Detection Network reported 30 million flashes each year. Globally, there are about 45 flashes per second.

In the United States there were about 60 deaths per year due to lightning strikes during 1979–2002. Lightning is also a major cause of forest fires. In the United States lightning starts an estimated 17 000 forest fires annually, with most occurring during June–August.

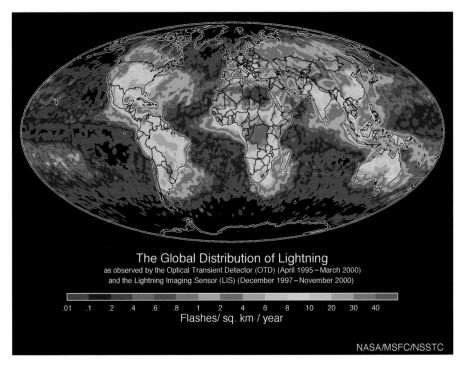

Figure 2.9 The global distribution of lightning flashes, 1995–2000, observed by satellite sensors. The Optical Transient Detector, April 1995–March 2000 and the Lightning Image Sensor, December 1997–November 2000 (source: NASA Marshall Space Flight Center).

(d) Precipitation

The term *precipitation* encompasses all hydrometeors (rain, snow, hail, dew, hoar frost, and rime). Globally, only about 5 percent of all precipitation falls as snow, due to the high rainfall totals in low latitudes.

Raindrops form by one of two mechanisms. The initial process involves droplets forming at the cloud condensation level on hygroscopic nuclei such as sea salts and sulfate particles. Because the saturation vapor pressure is less over a salt solution droplet than a pure water droplet, the solution droplets tend to grow preferentially. However, the saturation vapor pressure is also greater over a curved droplet than the flatter surface of a large drop, and hence the small droplets tend to evaporate. The droplets grow only slowly by vapor accretion (especially as more drops are needed as the drop diameter increases), but this process can be accelerated by the coalescence process. Slightly larger drops fall faster and sweep up smaller droplets in their wake as they fall, thus growing into raindrops. Rain drops range in diameter from about 0.5 to 5 mm, with corresponding terminal velocities (fall speed) of 2–9 m s^{-1}. A second mechanism is common over hills where there is a cap cloud of water droplets. In winter in middle latitudes, the presence of overlying mid-level cloud may release ice crystals that seed the lower feeder cloud. This is known as the seeder–feeder mechanism proposed by Tor Bergeron. The water droplets in the lower feeder cloud accrete onto the falling ice crystals because the saturation vapor pressure is lower over an ice surface than over a water surface. The maximum difference is about 0.2 mb at $-12\,°C$. The ice crystals then grow by collisions into snowflakes. These may reach the ground as snow, or rain if the freezing level is 250 m or more above the surface, so that the flakes melt as they fall through the warmer air below the cloud.

Rainfall is collected in a precipitation gauge of which there are about 80 different types in the world's national meteorological services. There are a great variety of designs and sizes. The standard gauge in the United States is 8 inches in diameter. There are some 4950 gauges across the United States and around 200 000 world-wide. There are a much smaller number of recording gauges that allow the rainfall rate to be measured. In windy locations gauges may have a windshield to smooth the airflow over the top of the gauge. The water equivalent of snowfall is obtained by slowly melting the snow in the gauge. In snowy climates a fence may be erected around the gauge to improve the catch. Snowfall totals from unshielded gauges in windy climates may be only half of the true amount falling on the ground. The World Meteorological Organization (WMO) recommends a double-fence design for accurate snowfall measurement. Because of different gauge types, different shielding, local exposure factors, and precipitation characteristics (type, intensity), precipitation data should be regarded as approximate values. Bias is introduced by the height of the gauge rim above the ground surface, with amounts decreasing as the rim is elevated. The undercatch due to this factor ranges from 2 to 16 percent in the United Kingdom. Estimates over ocean areas are now usually derived from satellite passive microwave data. Estimates of rainfall rate on land can be determined from weather radars that track the vertical and horizontal characteristics of clouds, the storm's movement, and rainfall intensity.

A precipitation index that is often used is the number of rain days above some threshold amount during 24 hours. However, the threshold is different between countries (0.25, 1, 2, or 5 mm), making global comparisons impossible. Rainfall intensity relates the amount to its duration; this value is indicative of flood risk, for example. Sometimes the mean intensity is determined per rain day. Annual mean amounts per rain day (> 1 mm) range from 5 mm in London to 14 mm in Lagos, Nigeria, and 22 mm at Mumbai, India.

In tropical climates, precipitation occurrence has a strong diurnal variation. Over land areas, maximum rainfall occurs in the late afternoon, but over oceans it occurs at night or in the early morning. The land maximum is related to the daytime surface heating and the build up of cumulonimbus clouds. The nocturnal maximum over the ocean is related to radiative cooling of the cloud tops at night that leads to a destabilization of the lower atmosphere, triggering convection.

Precipitation is the most widely varying climatic element in time and space due to its small-scale structure. Thunderstorm cells are typically 1–10 km in diameter and the associated rain and hail lasts an hour or so at a given location. Cyclonic low-pressure systems give rise to widespread precipitation, covering hundreds of kilometers, but there is small-scale cellular structure (10–50 km) within the overall precipitation area. In general, the average rainfall over a large area (100 000 km^2) is only 10–30 percent of that falling over a small area (25 000 km^2). As well as the area of a storm, we are also interested in precipitation intensity. Precipitation may fall as drizzle, continuous rainfall lasting hours, or showers of short (~ 1 hour) duration. World record values range from about 40 mm in a minute to 400 mm in an hour, and 1800 mm in a day. The rain intensity drops off sharply as the time interval considered is increased. Maximum average annual amounts reach ~ 12 m in a few mountainous locations (Lloró, Columbia; the Assam Hills, India; Mt. Waialeale, Hawaii; and the west slope of the Southern Alps, New Zealand). By contrast, desert areas of the world receive little or no precipitation in the same time interval.

The global distribution of annual and seasonal precipitation is complicated (Figure 2.10). In boreal winter the West Pacific "warm pool" of high sea surface temperatures, which sets up deep convection, is prominent and in boreal summer the Intertropical Convergence Zone (ITCZ) between the hemispheric trade wind systems across the North Pacific Ocean is evident (see Section 3.1b). The equatorial continental heat sources in South America, Africa, and the "maritime continent" of Indonesia and the Philippines also show up. Globally, some 90 percent of total evaporation is precipitated back to the oceans, and only 10 percent falls on land areas.

In tropical and subtropical cloud forest environments, fog drip onto leaves and branches is an important additional source of precipitation. It may add 20–30 percent to the direct rainfall amount. Trade wind cloud is one source of such moisture on the windward slopes of tropical islands in the eastern oceans. Typical locations for fog drip include the Canary Islands in the eastern North Atlantic and the Hawaiian Islands. In winter in mid-latitudes the fog may be at freezing temperatures and then rime ice is deposited on vertical surfaces.

Hail consists of nearly spherical or irregular lumps of ice. Hailstones are composed of alternating rings of clear-water ice and opaque granular ice measuring between 5 and 200 mm in diameter. They form in thunderstorm cumulonimbus clouds where there is a strong updraft allowing water droplets to accrete onto a nucleus of soft hail (graupel). The hailstone may fall through the cloud, causing melting, and then be swept up again by the moving storm, allowing the melted

(a)

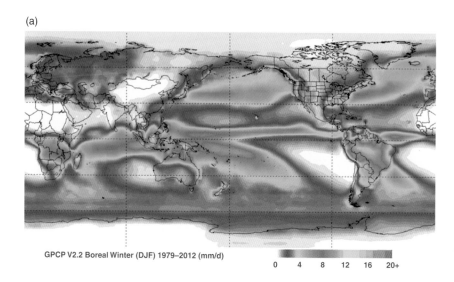

GPCP V2.2 Boreal Winter (DJF) 1979–2012 (mm/d)

0   4   8   12   16   20+

Figure 2.10
Seasonal maps of global precipitation, 1979–2012 from the Global Climatology Precipitation Project, using combined gauge and satellite data (mm day^{-1})
(a) December–February;
(b) March–May;
(c) June–August;
(d) September–November (source: NASA, GSFC, Mesoscale Atmospheric Processes Laboratory).

(b)

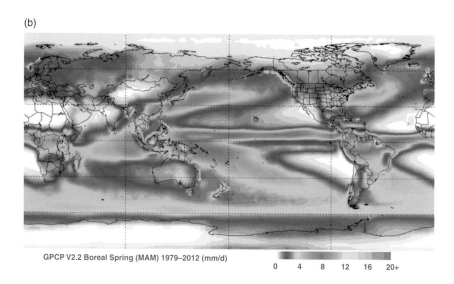

GPCP V2.2 Boreal Spring (MAM) 1979–2012 (mm/d)

0   4   8   12   16   20+

surface to refreeze above the freezing level. The hailstone collects super-cooled droplets that form an opaque ice layer. It is also possible that the alternating layers are related to variable conditions within the thundercloud and the fall-speed of the hailstone versus the updraft velocity.

Satellite data from the Advanced Microwave Scanning Radiometer (AMSR)-E show that large hail (> 2.5 cm diameter) is common in northern Argentina and Paraguay, the central United States, Bangladesh, Pakistan, central and West Africa, southeast Africa and eastern Asia. The largest hailstone recorded

Figure 2.10
(cont.)

(c)

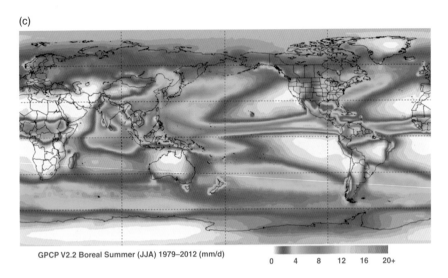

GPCP V2.2 Boreal Summer (JJA) 1979–2012 (mm/d)

0   4   8   12   16   20+

(d)

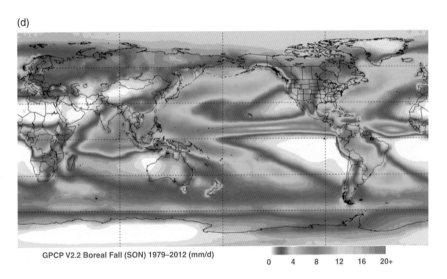

GPCP V2.2 Boreal Fall (SON) 1979–2012 (mm/d)

0   4   8   12   16   20+

in the United States fell on July 23, 2010 in Vivian, South Dakota; it measured 20 cm in diameter and weighed 0.87 kg. Hail is a major source of damage to crops, as well as causing property damage and risks to humans and animals exposed to it. Crop damage in the United States amounts to about US $1.3 billion annually; in the Great Plains the losses represent 5–6 percent of the crop value according to S. Chagnon.

Global *snow cover* can be described by the extent of snow-covered area (SCA) (Figure 2.11) and by the snow water equivalent (SWE) of the snow pack

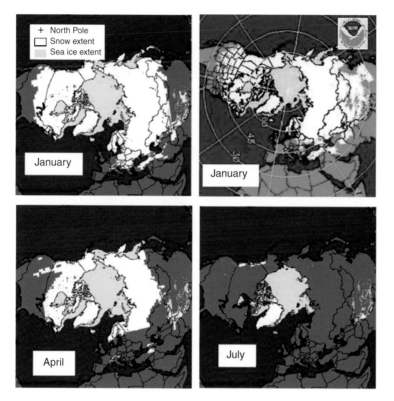

Figure 2.11
Extent of Northern Hemisphere snow cover (white) for January, April, and July, 1967–2005 (NSIDC). The top right panel shows state boundaries for January (NOAA). Sea ice (blue except top right where it is yellow) is shown for 1979–2003 (NSIDC).

(Figure 2.12) on particular dates. In January, the area with snow cover in the Northern Hemisphere averages 47 million km². Snow covers the ground for 7–8 months of the year in high northern latitudes and a few weeks to months in middle latitudes. Snow depth at a weather station is usually determined by inserting a ruler into the snow pack at several spots at 09:00 hours daily. In the Russian Federation it is the average depth at three fixed stakes. Average snow depths range from 50–100 cm on the Eurasian steppes to 30–40 cm on Arctic sea ice. The Upper Peninsula of Michigan may accumulate 500 cm annually due to lake effect snowstorms that develop when cold air flows over the unfrozen lake in autumn and early winter and picks up moisture, forming clouds and snow bands (see Section 4.3). The SWE is the depth of water that would be obtained by melting the snow pack, which has variable density (~ 100–400 kg m^{-3}). On average, snow pack density is about 300 kg m^{-3}. It is difficult to estimate SWE on continental scales due to the upper limit of passive microwave data (~ 0.8 m snow depth) and the effects on microwave emissivity of forest cover, frozen layers, and hoar frost layers in the snow pack.

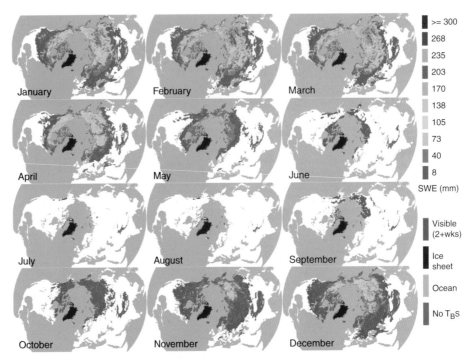

January   February   March

April   May   June

July   August   September

October   November   December

>= 300
268
235
203
170
138
105
73
40
8
SWE (mm)

Visible
(2+wks)

Ice
sheet

Ocean

No T_Bs

Figure 2.12 Mean Northern Hemisphere SWE by month from blended Special Sensor Microwave Imager (SSM/I) passive microwave and visible data T_B = brightness temperature (source: R. L. Armstrong and Mary Jo Brodzik, NSIDC).

(e) Evaporation

The *evaporation* of water from the ocean or a moist surface requires an energy source and wind to remove the evaporated water vapor. The rate of evaporation also depends on the vertical gradient of vapor pressure in the surface boundary layer. The latent heat of vaporization (required to convert liquid to vapor) is approximately $2.5 \times 10^6 \, \mathrm{J\,kg^{-1}}$ (see Appendix). The energy required to evaporate a given mass of liquid water corresponds to approximately 600 times the energy required to increase its temperature by 1 °C, and to 2400 times the energy required to increase the temperature of a corresponding mass of air by 1 °C. Methods to calculate evaporation use either an aerodynamic approach where the vertical gradients of wind speed and water vapor between about 1 and 10 m height are used, or the energy budget method where the energy required for evaporation is calculated. In 1948 H. L. Penman in England developed an approach that combined the aerodynamic calculations with those of the energy budget, and this is generally considered to be the most accurate formulation. An analogous method was prepared by R. O. Slatyer and

I. C. McIlroy in Australia in 1961. The original Penman equations used air temperature data, but W. J. Shuttleworth modified them in 1993 to use solar radiation.

Direct measurements are possible using one of three methods: evaporation pans where the change in water level in the pan is measured daily; a lysimeter, where a block of vegetated ground is weighed to measure changes due to evaporative losses; and eddy correlation systems, which record the instantaneous exchanges of moisture by vertical air currents near the surface. The evaporation pan is about 120 cm in diameter by 25 cm deep. Changes in the water depth can be scaled to estimate the loss from lakes and reservoirs. However, the pan has problems in dry environments where birds and animals drink from it. The lysimeter is useful for climatological estimates. Non-weighing lysimeters simply measure the drainage that percolated through the soil block, whereas weighing lysimeters measure precipitation, drainage and the change in water storage by the block. The lysimeter may range in size from a meter or so up to a plot 30–50 m across that contains a tree growing in it. If a lysimeter is regularly watered, the water loss is termed the potential evaporation – the maximum possible for the given climatic conditions. Knowledge of this quantity is useful in assessing irrigation requirements. Eddy correlation instruments typically use infrared spectrometers to measure water vapor and sonic anemometers for vertical airflow. The estimates of evapotranspiration by eddy correlation methods have been shown to underestimate the true amount by 20–30 percent.

The global FLUXNET program operates about 540 micro-meteorological tower sites around the world, of which ~ 400 are active; they use the eddy correlation method to determine fluxes of water vapor and carbon dioxide. They are installed in temperate conifer and broadleaved (deciduous and evergreen) forests, tropical and boreal forests, croplands, grasslands, chaparral, wetlands, and tundra. A tabulation of the sites and their data records is available at http://fluxnet.ornl. gov/site_status.

Vegetation transpires moisture through pores known as stomata. Hence, total moisture transfer from the surface is often referred to as evapotranspiration. The maximum possible evaporation from a wet surface is referred to as the potential evaporation. This occurs as long as there is a moisture supply, energy to evaporate it, and wind to remove the vapor.

Precipitation efficiency is the basis of a climatic classification system developed in 1948 by C. W. Thornthwaite. The system incorporated a moisture index, which relates the water demand by plants to the available precipitation by means of an index, calculated from measurements of air temperature adjusted for day length. In dry climates the moisture index is negative because precipitation is less than the evaporation.

The precipitation efficiency is based on the ratio of precipitation to evaporation calculated for each month and summed. The values are as follows:

A	Perhumid	>100
B_4	Humid	80 to 100
B_3	Humid	60 to 80
B_2	Humid	40 to 60
B_1	Humid	20 to 40
C_2	Moist subhumid	0 to 20
C_1	Dry subhumid	−20 to 0
D	Semi-arid	−40 to −20
E	Arid	−60 to −40

The system also uses an index of thermal efficiency, with accumulated monthly temperatures ranging from 0 (a frost climate) to greater than 127 (a tropical climate). The classification was applied for several continents, but no world map was ever produced.

Maps of evaporation indicate that in January there are maximum values of 300 mm off Japan and 250 mm off the eastern United States due to cold airflows over warm ocean currents (Figure 2.13). In July, maxima of ~200 mm are found over subtropical oceans in the Southern Hemisphere. In both seasons average maximum values over land are ~75 mm month^{-1} according to M. I. Budyko. Measurements over the Amazon basin indicate higher values of 110–130 mm month^{-1} from September to April and 80–90 mm month^{-1} during May–August.

On a global scale, about 85 percent of total evaporation is from the oceans and only 15 percent from land areas. The evaporation from land areas averages about 1.3–1.6 mm day^{-1}.

There is a *water balance* analogous to the radiation balance. The water balance equation is:

$$P = E + r \pm \Delta S,$$

where P = precipitation, E = evaporation, r = runoff, and ΔS = storage change in the soil or the snow pack. Each term may vary considerably over the course of the year if the precipitation is seasonal. Hence, there may be times when the soil dries out and there is no evaporation or runoff. Berkeley, California, illustrates a summer deficit after stored soil moisture is used (Figure 2.14), whereas Halifax, Nova Scotia, on the east coast of Canada, partially uses its stored soil moisture and then recharges it. The snow pack may store precipitation for up to 6–8 months and then release it via snowmelt in 1–2 months.

Drought is a critical water balance concept, but the term is hard to define on a universal basis. An absence of rainfall for a month or more that will go unnoticed in semi-arid environments may have serious consequences for agriculture in

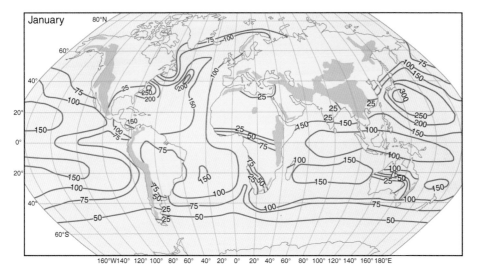

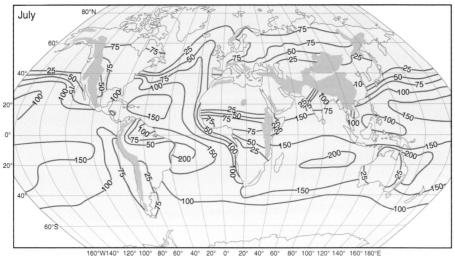

Figure 2.13
Maps of global evaporation (mm) in (top) January and (bottom) July (source: after Budyko 1958, from Barry and Chorley 2010).

mid-latitudes. Meteorological drought is defined in terms of rainfall anomalies, and agricultural drought in terms of soil moisture deficit, while hydrological drought relates to runoff conditions. Obviously, meteorological conditions are the first determinant, followed by soil moisture, and then stream flow. C. W. Thornthwaite devised a monthly book-keeping approach to the water balance, keeping track of the inputs and outputs. He estimated evaporation based on air temperature. In 1966 Wayne Palmer developed the Palmer Drought Severity Index (PDSI), which also uses temperature and precipitation data. Zero is normal and −4 is extreme drought; +4 is excessive rainfall. Droughts have many causes, but a principal factor is the presence of persistent high-pressure conditions that promote subsiding air and clear skies leading to dry weather. Once soils

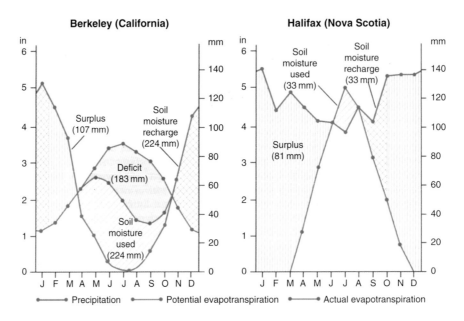

Figure 2.14 The annual cycle of water balance at (left) Berkeley, California and (right) Halifax, Nova Scotia (based on Thornthwaite and Mather 1955; from Barry and Chorley 2010).

become dry there is a feedback effect because there is then no local evaporation. In 1968 J. Keetch and G. Byram devised a drought index (KBDI) specifically to assess wildfire potential. KBDI is an index of soil/duff drought that ranges from 0 (no drought) to 800 (extreme drought), based on a soil capacity of 200 mm (8 inches) of water and hundredths of an inch of soil moisture depletion, calculated on the basis of a daily water balance. The depth of soil needed to hold 200 mm of water varies. Real-time maps of KBDI for Florida and Texas are available online.

In 1955 M. I. Budyko in the Soviet Union proposed a way of combining the energy and water balances in a single index termed the radiational index of dryness. He related the net radiation available for evaporation from a wet surface (Ro) to the heat required to evaporate the mean annual precipitation (r). The ratio Ro : Lr, where L = latent heat of evaporation, is the radiational index of dryness. It has threshold values for climate/vegetation zones as follows: > 3 = desert; 2–3 = semidesert; 1–2 = steppe or prairie; 0.33–1 = forest; and < 0.33 = tundra. Unlike Köppen's classification, which uses arbitrary thresholds of temperature and precipitation based on empirical associations with vegetation (see Figure 1.3), Budyko provided a physically based relationship. Analysis of data from 167 FLUXNET sites around the world shows that the Budyko relationship explains 62 percent of the variation in evaporative index – the fraction of mean annual precipitation used in evapotranspiration. An additional 13 percent is accounted for by vegetation differences.

(f) Visibility

Visibility is defined as the distance at which an object is visible; at night it is defined with respect to a light source. The visibility reported at weather stations is that prevailing in at least half of the horizon circle. It is a critical measure for many modes of transportation – highways, railroads, shipping, and aircraft. It is affected by the current weather, including fog, haze, rain, snow, and blowing snow or sand. The lowest category is < 50 m visibility. Obscuration of visibility is caused by suspended water droplets (fog), or ice crystals (ice fog), by aerosols (haze), by falling rain or snow, and by blowing snow and blowing sand. Fog is said to occur when the visibility is < 1 km. Fog forms when the relative humidity close to the ground is near 100 percent and there is a light wind to mix the moisture upward. The air achieves saturation either by cooling the air to its dew-point temperature, or by adding moisture into the air by evaporation. Radiation fog forms by cooling, whereas advection fog forms when warm air is cooled as it flows over a colder surface. Steam fog occurs when moisture condenses above a warm water body, as is often the case over high latitude oceans in winter. Haze is said to occur when dust or smoke or other particles in suspension lower the visibility. A well-known example is the "Asian brown cloud" over southern Asia, mainly during the dry winter monsoon (December–April). The source of the haze is pollution from wood burning, biomass burning, industry, and transportation.

Another regional phenomenon is Arctic haze, which is a result of the poleward transport of industrial pollution (mainly sulfur) and Asian dust particles. It is persistent in winter and early spring due to the stability of the Arctic atmosphere and the absence of precipitation to wash it out.

Blowing snow occurs when snow on the ground is lifted by the wind above 2 m height; this requires wind speeds above 12–15 m s^{-1}. It can also occur when snow is falling with a strong wind. It is a common feature of the North American prairies or Russian steppe in winter, as well as the polar regions.

Blowing sand is common in arid and semi-arid regions. Strong winds, associated often with an advancing weather front (see Section 3.3) can raise sand particles into the air up to 1 km or more. Such a sandstorm is known as a haboob.

SUMMARY

This chapter surveyed the climatological elements related to energy – the radiative components – temperature, and moisture conditions – clouds, precipitation, and evaporation. Some key points are noted. Solar (shortwave) radiation is reflected by clouds and the surface, giving a total planetary albedo of 30 percent. Terrestrial (longwave) radiation is emitted by the surface and much of this is absorbed in the atmosphere by water vapor and other greenhouse gases (carbon dioxide, methane). This radiation is re-emitted back to the ground and out to space. The atmosphere cools radiatively, but is warmed by sensible heat and latent heat transferred from the surface. Temperature is a measure of the heat contained in the air or ground. The environmental lapse rate averages about 6.5 °C km^{-1} in the troposphere. The atmospheric moisture content can be expressed by the vapor content, the vapor pressure, or the precipitable water. Precipitation includes all forms of hydrometeors, but globally most falls as rain. It may occur as showers or long-lasting storm system precipitation. Orography greatly modifies total amounts over complex terrain. World maximum amounts average around 12 m per year. In January, just under 50 percent of the land surface in the Northern Hemisphere is snow covered. Snow cover is a major component of the climate system due to its high albedo and insulating effect on the ground beneath it. Evaporation transfers moisture from the surface to the atmosphere. It requires an energy source and wind to remover the evaporated water. Drought occurs when evaporation over land exceeds rainfall for a lengthy time interval.

QUESTIONS

1   The layer of the atmosphere where most of the world's weather occurs is:
    (a) thermosphere
    (b) mesosphere
    (c) stratosphere
    (d) troposphere
    (e) hemisphere.
2   The tropopause is highest at the North and South Poles. True or false?
3   Contrast the vertical gradients of temperature, pressure, and vapor content.
4   Compare the exchanges of solar radiation and terrestrial infrared at the surface.
5   What is meant by the "Greenhouse Effect"?
6   Discuss the sources of error in precipitation measurements.
7   Compare the different methods of measuring evaporation and examine their limitations.
8   The Bowen ratio is >1 when the surface is dry. True or false?
9   Over a dry surface LE > H. True or false?
10 G is minor compared to H. True or false?

The elements of climate

A global view of pressure, winds, and storms

3

Hurricane Isabel (2003) seen from the International Space Station. The eye, eyewall, and surrounding rainbands are clearly visible.

THIS CHAPTER continues our examination of climatic elements, treating air pressure, winds, air masses, frontal zones, storm tracks, thunderstorms, and mesoscale convective systems. These can be considered as the dynamic elements of the climate system. Pressure differences are the driving force for winds and storm tracks. Air mass contrasts give rise to frontal zones and to changing weather conditions. Thunderstorms may occur as single-cell or multicell clusters. Mesoscale systems are usually violent, moving storms, intermediate in size between thunderstorms and synoptic cyclones. The latter are low-pressure systems that are featured on a weather chart, with a diameter of 1500–5000 km and a lifetime of 5–7 days (see Section 3.4).

3.1 Pressure and winds

(a) Pressure

Atmospheric pressure is the force per unit area exerted on a surface by the weight of air above it. Pressure decreases with altitude as there is less overlying weight of air. Observations of pressure at a station, made by measuring the height of a mercury column in a mercury barometer, are "reduced" to pressure at mean sea level (MSL) by corrections for the latitudinal variation of gravity and for air temperature. At MSL the mercury column that is supported by the weight of the atmosphere is approximately 760 mm (29.9 inches) tall. Atmospheric pressure at MSL averages about 1013 millibars (mb) (or 1013 hPa, hectopascals – the pascal unit of pressure equals a force of one newton per square meter. In this text we use the traditional millibar unit). The record low value was 870 mb, in the eye of super-typhoon Tip in the northwest Pacific in October 1979. The highest recorded MSL pressure was 1083 mb, measured at Agata, in northeastern Russia in December 1968. The daily or weekly change of pressure is recorded by a barograph. This is an aneroid barometer comprising an evacuated metal chamber that responds to air pressure changes. It is connected via a series of levers to a pen that traces the pressure on a chart attached to a rotating drum.

The spatial pattern of pressure is largely zonal, with low values near the Equator (the equatorial trough of low pressure), flanked in each hemisphere by subtropical high-pressure cells (anticyclones), which are in turn flanked by subpolar low pressure (see Figure 3.1). The subtropical anticyclones extend vertically through the troposphere and are present throughout the year. The Northern Hemisphere anticyclones at the surface are known as the Azores high pressure, and the North Pacific high (Figure 3.2). The Southern Hemisphere anticyclones are correspondingly located in the southeast Pacific, the southeast Indian Ocean and southeast Atlantic oceans (Figure 3.3). In winter in the Northern Hemisphere there is also a pronounced, but shallow, high-pressure cell over northeastern Siberia, related to cold, dense air (Figure 3.2a). The high-pressure cells shift very little on a day-to-day basis. The subpolar lows in the Northern Hemisphere are located near Iceland and the Aleutians in winter, while in summer the former weakens and the latter disappears (Figure 3.2b). In the Southern Hemisphere there is a more or less continuous subpolar low around Antarctica in both seasons (Figure 3.3).

On a weekly time scale, low-pressure systems (cyclones) move broadly eastward across the mid-latitude oceans and continents at about 20–30 km hr^{-1}. For a mature low-pressure system in mid-latitudes, the central pressure is about 960–990 mb. Lows form, deepen, and then decay. Anticyclones are larger in extent and much more static components of the weather map. In the Subtropics they may move only on a seasonal time scale. Their central pressure is typically around 1030–1040 mb.

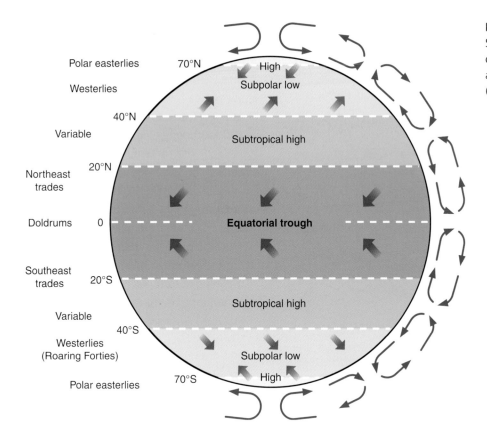

Polar easterlies

Westerlies

Variable

Northeast
trades

Doldrums

Southeast
trades

Variable

Westerlies
(Roaring Forties)

Polar easterlies

70°N

40°N

20°N

0

20°S

40°S

70°S

High

Subpolar low

Subtropical high

Equatorial trough

Subtropical high

Subpolar low

High

Figure 3.1
Schematic
diagram of wind
and pressure belts
(source: NASA).

In the troposphere and stratosphere the *geopotential height* of a pressure surface has been used for convenience since 1946, rather than the pressure values at a constant height. The 500 mb pressure surface is at a height of about 5.5 km (5500 geopotential meters, or gpm). The geopotential height is given by an adjustment to geometric height above MSL using the variation of gravity with latitude and elevation. Geopotential and geometric height are numerically almost equal.

Pressure decreases exponentially with altitude, whereas temperature decreases nearly linearly. Hence the pressure is about 1000 mb at sea level, 500 mb at 5.5 km, 100 mb at 15.5 km, and 25 mb at 25 km. The upward pressure gradient in the atmosphere is counter-balanced by downward gravitational acceleration. This gives rise to a state of hydrostatic equilibrium.

The change of pressure (p) with height (z), where ∂ is a small increment, is given by the air density (ρ, rho) times gravity (g), which is written:

$$\frac{\partial p}{\partial z} = -g\rho.$$

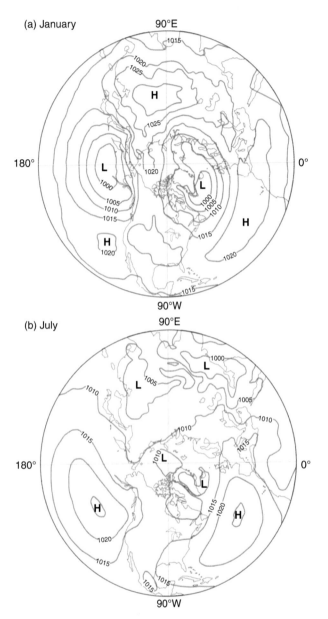

Figure 3.2 Mean sea level pressure (mb) in the Northern Hemisphere in (a) January and (b) July (source: Barry and Chorley 2010).

For reference purposes, a series of Standard Atmospheres have been developed that relate air pressure, temperature, and density at different heights above sea level at different latitudes. At sea level the Standard Atmosphere in mid-latitudes has a pressure of 1013.25 mb, a temperature of 15 °C, a density of 1.225 kg m^{-3}, and an initial lapse rate of -6.5 °C km^{-1}. At 5 km the pressure is 540.5 mb, the temperature is -17.5 °C and the density is 0.786 kg m^{-3}. The tabulation continues to 11 km, where the pressure has fallen to 226.3 mb and the temperature to -56.5 °C. Between 11 km and 20 km the temperature in the Standard Atmosphere remains constant.

(b) Winds

Air motion is given its initial impulse by horizontal pressure differences that result from contrasts in heating and air density. The wind flows towards lower pressure to transfer mass. *Wind velocity* is a vector quantity comprising the wind direction (360° is a north wind, 90° an east wind, and so on), referring to the direction from which the wind is blowing, and the wind speed at a standard height of 10 m above the ground, averaged over a minute or so. Wind speed is measured using an anemometer (*anemos* is Greek for wind). The anemometer was invented in 1848 by J. T. R. Robinson at Armagh Observatory in Ireland. It consisted of four hemispheric cups mounted on a rotating vertical spindle; modern versions have three conical cups set at 120° apart. Nowadays, sonic anemometers which measure ultrasound waves, invented in the 1950s, are often used, especially on automatic weather stations. Direction is measured by a wind vane to tens of degrees. Speeds are measured in m s^{-1}, the distance the air moves in one second. At some locations a continuous record of wind speed and direction is obtained with a pressure tube anemograph which records wind speed with a pitot device. Gale-force winds have speeds >17 m s^{-1}. Gusts are defined by brief (<20 seconds) increases of speed exceeding 8 m s^{-1}. The world-record gust is 113 m s^{-1} (253 mph), measured in a tropical cyclone off Western Australia. Squalls are like gusts, but typically lasting longer than two minutes. Above the surface, winds are determined by radar tracking of helium-filled sounding balloons called RAWINSONDES (radar wind sondes). Such soundings are made once or twice per day at about 700 stations around the world.

Wind data are sometimes displayed on a wind rose, which shows the frequency of winds from each direction, and their speed distribution in a polar coordinate plot. An example for La Guardia, New York, in 2008 is shown in Figure 3.4. This enables the most frequent wind directions to be easily seen. The resultant wind is the vector average of all directions and speeds. The mean wind speed is averaged for a day or month for all wind directions and calm conditions. In general circulation studies it is common to determine zonal westerly (u) and meridional southerly (v) components of the wind. Easterly winds are negative u and northerly winds are negative v.

The surface winds are basically organized in two extensive systems in each hemisphere – the tropical easterly *trade winds* and the mid-latitude westerly winds (Figure 3.1). The trade winds are highly constant in direction, whereas the *westerlies* are much more variable in direction

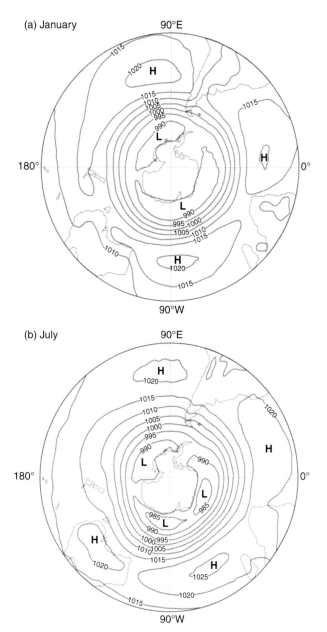

Figure 3.3 Mean sea level pressure (mb) in the Southern Hemisphere in (a) January and (b) July (source: Barry and Chorley 2010).

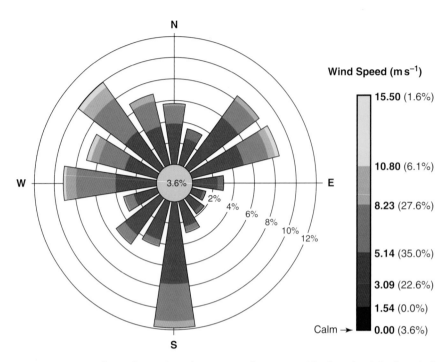

Figure 3.4 A wind rose for La Guardia, New York, in 2008. The length of the bars indicates the percentage frequency of a wind of given speed and direction; the value in the center is the percentage of calms (source: Wikipedia, http://en.wikipedia.org/wiki/Wind_rose).

and in speed, especially in the Northern Hemisphere, as a result of the presence of the land masses.

The winds above the friction layer ($\sim$ 1000 m deep) blow parallel to the isobars (lines of equal pressure), with low pressure on the left (right) in the Northern (Southern) Hemisphere. This is known as Buys Ballot's law. These so-called *geostrophic winds* represent a balance between the pressure gradient force towards low pressure and the Coriolis force due to the Earth's rotation acting in the opposite direction (see Box 3A.1). They have a velocity inversely proportional to the spacing between the isobars – closer spacing means stronger winds. Geostrophic winds describe the straight flow between parallel isobars. Near the surface, friction between the moving air and the surface slows down the flow, and hence the pressure gradient force exceeds the Coriolis force and causes the wind to turn toward lower pressure at an angle of about 15°.

When the isobars are curved, an additional force needs to be considered – centripetal force directed toward the center of the rotation. In cyclonic circulations the centripetal acceleration is toward the low, adding to the pressure gradient force, and resulting in the balanced flow, known as the gradient wind, being slightly less than the geostrophic wind. In anticyclonic circulations, the opposite

Box 3A.1 **Coriolis force**

The Coriolis force is often a source of confusion and misunderstanding. It is named after a nineteenth-century French scientist, C.-G. Coriolis. It is an artifact that arises as a result of the anticlockwise rotation of the Earth about its axis. Air motion over the Earth's surface is viewed with respect to the rotating reference frame of the Earth (that is, the latitude–longitude coordinate system) and hence motion is apparently deflected to the right (left) in the Northern (Southern) Hemisphere. In the seventeenth century it was already recognized that a cannon ball fired to the north would turn to the east due to the rotation of the Earth under it. The deflection is zero at the Equator, where the motion is parallel to the rotational axis, and increases with the sine of the latitude to a maximum at the Poles, where the motion is perpendicular to the axis. It is proportional to the rotation rate of the Earth and to the wind speed. It affects the wind direction but not the wind speed and corresponds to a force at right angles to the wind direction. The Coriolis force balances the pressure gradient force and gives rise to the geostrophic wind. At the Equator, where there is no Coriolis effect, air flows down the pressure gradient toward low pressure. The Coriolis force is operative for air motion on a large scale (tens of kilometers) and long time intervals (a day). The Coriolis force affects ocean currents and hence the Gulf Stream turns to the right as it crosses the North Atlantic Ocean.

Box 3A.2 **The thermal wind relationship and thickness**

Temperatures in the troposphere decrease from Equator to Pole and this gradient is paralleled by pressure surfaces. As a consequence there is a component of wind that blows from the west at right angles to the Equator–Pole pressure gradient. Standing with one's back to this wind, cold air is on the left in the Northern Hemisphere. This is known as the thermal wind and is a major component of the actual zonal (westerly) wind. The thermal wind is a vertical shear (change in speed or direction) in the geostrophic wind (which blows parallel to the isobars) that is caused by a horizontal temperature gradient.

The warm air column in low latitudes is deeper than a cold air column at high latitudes, because the warm air expands vertically, and this greater atmospheric thickness determines the strength of the thermal wind. As the thickness gradient changes seasonally, so the geostrophic wind strengthens from summer to winter.

The thickness is the depth of air between two pressure surfaces, usually 1000 and 500 mb (or hPa). It is proportional to the mean temperature of the layer – lower thickness implies cold air. The thermal wind blows parallel to lines of equal thickness with cold air to the left in the Northern Hemisphere. Extratropical cyclones tend to be steered parallel to the thermal wind.

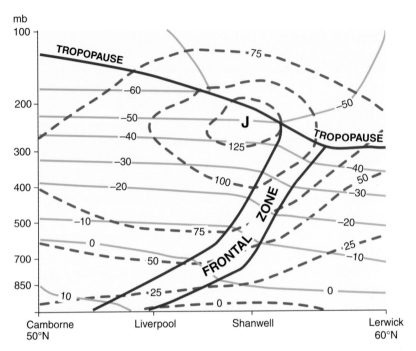

Figure 3.5 Cross-section of the atmosphere at 50–60° N through Great Britain, 26 September 1963. Full lines are isotherms °C; dashed lines isotachs of the westerly wind component in knots; 1 knot = 0.51 m s^{-1} (source: Barry 1967).

applies and the gradient wind is greater than in the geostrophic case. For most purposes, the geostrophic wind is a sufficient approximation to the gradient wind.

Above the surface, in the troposphere, there is a deep vortex centered near the Pole in each hemisphere. The vortex is circled by strong westerly winds. These arise from the Equator–Pole temperature gradient as a result of the thermal wind relationship (Box 3A.2). As a consequence, the upper winds are overwhelmingly westerly.

In the upper troposphere around 250–300 mb (8–10 km above sea level), there are westerly *jet streams* (narrow bands of high wind velocity, 40–65 m s^{-1}) in middle latitudes of each hemisphere. They were discovered by bomber pilots trying to fly westward across the North Pacific during World War II. Jet streams result from a concentration of the meridional temperature gradient in mid-latitudes, primarily between low-latitude tropical and high-latitude polar air. This is the polar front and the associated jet stream (Figure 3.5). This figure shows a cross-section of the troposphere from 50–60° N across Great Britain. There is also a subtropical jet stream caused by a meridional temperature gradient in the upper tropical troposphere. Figure 3.6 illustrates their circumpolar patterns and shows that in some longitudes both jet streams are present. They have a vertical depth of a few kilometers and a lateral extent of a few tens of kilometers. In the along-stream direction

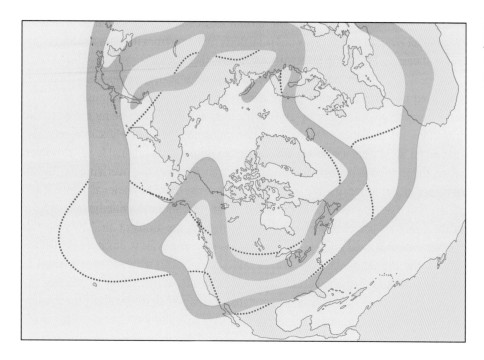

Figure 3.6
Jet streams on 27 November, 1958 (source: Riehl 1962).

there may be jet streaks – zones of higher wind speeds – extending several hundred kilometers. Sometimes jet streams are marked by bands of cirrus cloud that are visible on satellite imagery. On the equatorward side there is often a zone of clear air turbulence as a result of strong lateral wind shear.

The highest mean speeds (up to 135 m s^{-1}) are found over eastern Asia in winter as a result of the high constancy in location of the jet stream core. This is the location of a strong cold land–warm ocean temperature contrast. In the summer Tropics at about 15° N over India, extending into Africa, there is a high-level tropical easterly jet stream located at 150–200 mb. This reflects the north–south temperature gradient in the upper troposphere, caused by heating over the Tibetan Plateau, which regionally reverses the usual Equator–Pole temperature gradient. In winter the westerly subtropical jet stream is anchored south of the Himalaya, Tibet, with a variable polar front jet stream located much further north over Russia.

In some locations, low-level "jets" (LLJs) are observed, but these have typical speeds of only 15–20 m s^{-1}. Some are formed on the upwind side of mountain ranges, such as where an airflow toward the range is blocked because it lacks the kinetic energy required to cross the mountains. The westerly flow encountering the Sierra Nevada in California in winter often turns northward and forms a southerly jet at about 850 mb along the side of the range because there is a stable atmospheric layer over the mountains that prevents the air from rising. In spring and summer there is an important nocturnal LLJ over the High Plains of the United States that is related to the eastward slope of the terrain and cool, high-elevation air overlying warmer air over the plains. This LLJ transports moisture from the Gulf of Mexico into the Midwest, contributing greatly to severe weather events.

Vertical air motion happens on small and large scales. Small-scale motion is produced by surface heating giving rise to upward air motion through buoyancy differences (convection), or by air flowing over small hills and being forced to rise (mechanical turbulence). On average, the atmosphere is relatively stable, meaning that when air is displaced vertically it tends to return to its original level. However, when the ground is heated, and this heat is transferred to the overlying air, the air becomes unstable because the heated air is less dense than the cooler air above it and so tends to rise. The rising air cools and this cooling continues until the temperature of the rising air parcel comes into equilibrium with the ambient air. The rising air parcels cool at a fixed rate of 9.8 °C km^{-1}, as long as the air is unsaturated. This is termed the *dry adiabatic lapse rate* (DALR) because no external heat is used or supplied. Sinking dry air warms at the same DALR. When rising air reaches the condensation level, latent heat of condensation is released, lowering the cooling rate – this *saturated adiabatic lapse rate* (SALR) is typically 5–6 °C km^{-1}. This rate is not reversible, as when moist air starts to descend the air becomes unsaturated.

Larger-scale upward motion results when an airstream encounters a mountain range and has sufficient energy to rise over it (rather than being blocked or forced around). On the lee side of a mountain range, air that has flowed over it sinks due to gravity. Large-scale vertical motion also occurs in low- and high-pressure areas. Air flows inward toward a low-pressure center at the surface as a result of slowing of the motion due to friction at the surface; this allows the pressure gradient force to exceed the Coriolis deflecting force, turning the flow inward. The air converges in the center of the low and is forced to rise (Figure 3.7). In the upper troposphere above 500 mb this air diverges (flows outward). Conversely, in a high-pressure system, air sinks and flows outward at the surface. The sinking air is a result of airstream convergence at upper levels. The sinking air gives rise to clear skies, whereas the low pressure ascent forms clouds and rainfall. Small-scale vertical motion in thunderstorm updrafts can have velocities of 10–30 m s^{-1}, while large-scale vertical motion is only a few centimeters per second and is not directly measurable.

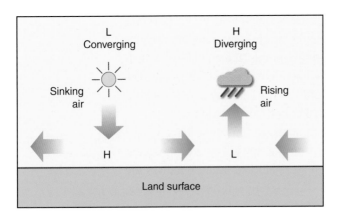

Figure 3.7 Schematic illustration of the vertical motion in high- and low-pressure centers and the associated patterns of low-level and high-level convergence and divergence of air.

(c) Horizontal moisture flux

The transport of moisture in the atmosphere is concentrated around the 900–700 mb level (1–3 km), where vapor content is still high and the wind speed has increased with height above the surface. Sounding balloons measure wind and vapor content at multiple levels in the atmosphere.

In January there are intense *tropospheric "rivers"* (bands of atmospheric moisture) that flow from southwest to northeast over the North Atlantic and North Pacific, and from northwest to southeast over the South Atlantic, South Pacific, and South Indian Ocean, as shown by Newell and Zhu. In July, there is additionally a strong feature over Southeast Asia. Filtering vapor flux vectors to remove time variations longer than three days allowed the moisture filaments to be identified. Four or five of these rivers account for most of the meridional moisture transport in the atmosphere, according to research by Zhu and Newell. The moisture bands were found to be related to the locations of atmospheric polar fronts, and in particular to warm cyclonic conveyor belts (see Section 3.3). The depth of the rivers is about 4 km and their width ∼400 km (Figure 3.8), based on studies over

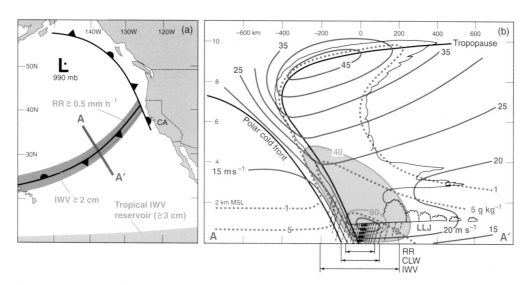

Figure 3.8 Conceptual representation of an atmospheric river over the northeastern Pacific Ocean (source: Ralph *et al.* 2004). (a) Plan-view schematic of concentrated integrated water vapor (IWV) (≥2 cm; dark green) and associated rain-rate enhancement (RR ≥ 0.5 mm h^{-1}; red) along a polar cold front. The tropical IWV reservoir (>3 cm; light green) is also shown. The bold line AA′ is a cross-section projection for (b). (b) Cross-section schematic through an atmospheric river (along AA′ in (a)) highlighting the vertical structure of the along-front isotachs of wind speed (blue contours; m s^{-1}), water vapor specific humidity (dotted green contours; g kg^{-1}) and horizontal along-front moisture flux (red contours and shading; ×10^5 kg s^{-1}). Schematic clouds and precipitation are also shown, as are the locations of the mean width scales of the 75 percent cumulative fraction of perturbation IWV (widest), cloud liquid water (CLW), and RR (narrowest) across the 1500 km cross-section baseline (bottom).

Figure 3.9
The global
sources and sinks
of moisture for
(top) DJF and
(bottom) JJA,
1980–2012
(source: Gimeno
et al. 2012).

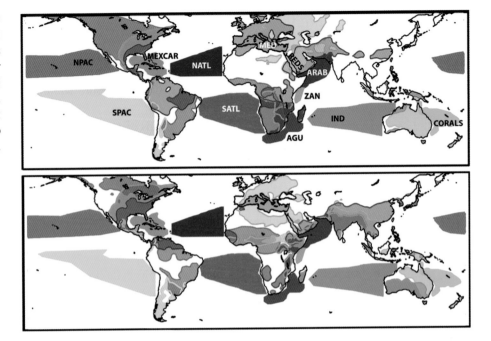

the North Pacific. However, within this band most of the variation in cloud liquid water and rainfall rate is concentrated within a core only 150–175 km wide. A study of the ten largest winter floods on the River Eden, Cumbria, UK between 1970 and 2010 shows that all of them occurred with southwesterly atmospheric rivers, associated with large-scale ascent of air and orographic uplift.

On a planetary scale, there is a net zonally averaged moisture flux (equal to precipitation minus evaporation) directed equatorward in the Tropics, as a consequence of the easterly trade wind systems, and poleward in middle latitudes of each hemisphere due to the westerly wind belts.

Evaporation exceeds precipitation in the subtropical anticyclone belts. The regional distribution of moisture flux has been analyzed through back trajectory analysis for 1958–2001 by Gimeno *et al.* A global summary for June–July–August (JJA) and December–January–February (DJF) is presented in Figure 3.9. It shows that the North Atlantic provides moisture for precipitation in eastern North America, Central America and northern South America during JJA, while in DJF it also provides moisture to Europe, North Africa and central South America. The western Pacific warm pool provides moisture for precipitation over Central and northern South America. The North Pacific and South Pacific sources provide moisture for precipitation over the latitudinal extremes of North and South

America, particularly in the respective hemispheric winters. The sources are numerous for the Indian Ocean region. In DJF four sources (the largest spanning most of the ocean between Australia and South Africa, one over the Agulhas current (see Figure 8.1), one over the Zanzibar Current, and one in the Arabian Sea) provide moisture for surrounding continental areas of the African continent and the Arabian Peninsula. In JJA the monsoon circulations make these four areas into moisture sources for precipitation over the Indian Peninsula. In DJF the Mediterranean and the Red Sea provide moisture for areas to their northeast. Finally, during JJA there are sources over the Gulf of Mexico for the southern and eastern United States, in the Coral Sea, and over subtropical South America. In general, the patterns reflect the role of warm conveyor belts (see Section 3.3) over the mid-latitude oceans and the systems of low-level jets in the Americas east of the cordilleras (e.g., the southerly jet over the High Plains of the United States in summer).

3.2 Air masses

An *air mass* is a large volume of air with particular temperature and moisture conditions that tend to be relatively homogeneous in the horizontal direction over a wide area. Tor Bergeron in Sweden developed the concept in the 1920s. At low levels in the atmosphere an air mass tends to take on the characteristics of its underlying surface – snow cover, warm/cold ocean, moist/dry and hot/cold land. The primary source regions for air masses, shown in Figure 3.10, are the snow-covered northern continents in winter (continental Polar or cP air), the subtropical oceans (maritime Tropical or mT air), and the tropical deserts (continental Tropical or cT air). Near the ground cP air is cold, dry, and stable, cT air is hot, dry, and unstable, while mT air is warm, moist, and stable, as it is moving over colder surfaces.

For some purposes, Arctic air (mA and cA) may be defined in high northern latitudes in winter, with a corresponding Antarctic (AA) type in high southern latitudes. Continental Arctic air is very cold, dry, and stable in winter in its source region over the Arctic Ocean. In summer it is replaced by maritime Arctic air due to the melting snow and ice surface and, increasingly, the greater open water area due to shrinking sea ice cover. Near the Equator the ocean air mass is designated as Equatorial maritime (mE), which is hot and humid with generally unstable air.

The mid-latitude oceans are indicated to have maritime Polar (or mP) air in Figure 3.10. This is a secondary air mass formed by the modification of cP air as it flows out of the continent over a warmer ocean surface and evaporates water from the surface, becoming moist and unstable due to surface heating. Notice that cP, cT, and mT air masses all form in areas of high pressure where the air movement is slow, so that the air mass has at least 5–7 days to develop its characteristics. As air masses travel away from their source region they tend to lose their distinguishing characteristics.

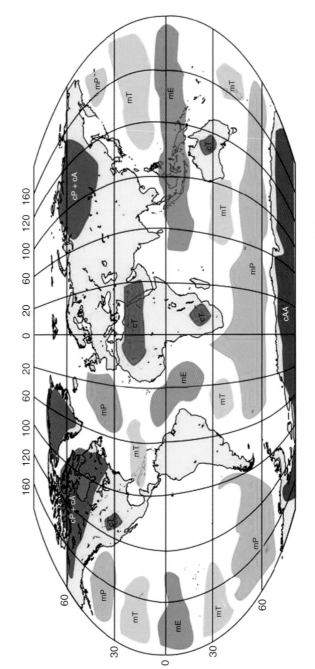

Figure 3.10 Air mass source regions (source: Wikipedia, http://en.wikipedia.org/wiki/File:Air-masses-.svg).

3.3 Frontal zones

Frontal zones form between contrasting air masses – especially polar and tropical air. The term "front" was adopted in Norway during World War I as an analogy to the battle fronts in Europe. Airstreams with different characteristics converge along an extensive boundary or frontal zone. The primary zone is the polar front between mT and cP or mP air. The polar front is located in middle latitudes shifting between about 30° N in the North Atlantic in winter and 60° N in summer (Figure 3.11). In summer there is an additional Arctic frontal zone located near

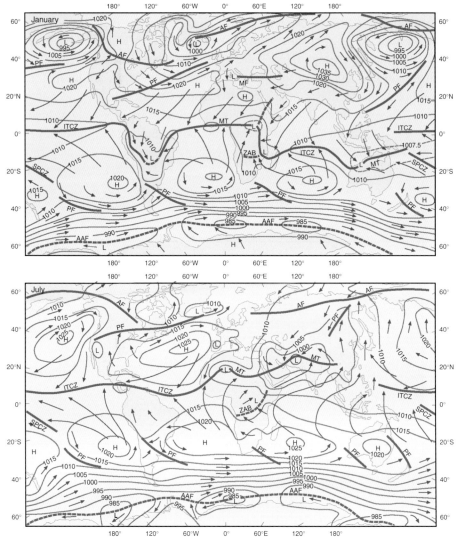

Figure 3.11
Frontal zones and MSL isobars of pressure (mb) in (January) winter and (July) summer.
AF = Arctic Front; PF = Polar Front (source: (Barry and Chorley 2010).

the Arctic coastlines of Eurasia and Alaska. It is attributed to the thermal contrast between the cold Arctic Ocean, overlain by shallow, cold Arctic air, and the snow-free tundra overlain by warmer cP air, but differences in energy budget between the tundra and boreal forest may be a supplementary factor. In the Southern Hemisphere the polar front is more zonal and represents the transition zone between cold polar easterlies and mild mid-latitude westerlies.

Frontal zones are also known as baroclinic zones, where pressure and density surfaces intersect in a vertical cross-section of the atmosphere. This is in contrast to air masses, which are essentially barotropic, where pressure and density surfaces are parallel to one another with height. Baroclinic zones give rise to jet streams in the upper troposphere (see Section 3.1b, Figure 3.5) and these tend to steer the movement of low-pressure systems.

Figure 3.12 shows atmospheric fronts over the globe as identified by an objective analysis as the average of four independent reanalysis data sets for 1989–2009. High frontal frequencies, of the order 10 percent of the time, are located in the extratropical storm tracks in the North Atlantic, North Pacific, and Southern Oceans. There are high frequencies close to high terrain, especially at high latitudes (e.g., east of the Canadian Rocky Mountains). Subtropical frequency maxima are evident near the South Pacific Convergence Zone (SPCZ) and near Taiwan in the western Pacific.

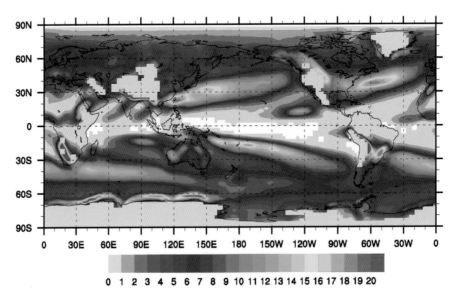

Figure 3.12 Annual mean frontal frequency (percentage of analysis time) for the period 1989–2009, averaged over four reanalysis data sets (source: Berry *et al.* 2011).

Analysis of rainfall associated with fronts shows that in the Southern Hemisphere 20 percent of rainfall is associated with both warm and cold fronts and 11 percent with quasi-stationary fronts. Corresponding values in the Northern Hemisphere are 17, 14, and 11 percent, respectively. Precipitation over the oceanic storm tracks is mostly associated with cold fronts – up to 42 percent for the North Atlantic – while over the Northern Hemisphere continents it is mainly associated with warm fronts – 30–58 percent over North America. In the Northern Hemisphere there are strong seasonal variations, with smaller percentages in summer; the Southern Hemisphere has much fewer seasonal variations. For mid-latitudes, 30–60° S, the figures are 29, 28, and 11 percent for warm, cold, and quasi-stationary fronts, respectively.

Between the northeasterly trade wind system of the Northern Hemisphere and the southeasterly system of the Southern Hemisphere there is a zone of convergence known as the *Intertropical Convergence Zone* (ITCZ). Convergence has two components – the confluence of two merging airstreams and speed convergence where the flow slows down and so air piles up. These two components may reinforce one another or cancel each other out. Similarly, divergence may consist of flow difluence and/or speed divergence where flow accelerates downstream. Convergence (divergence) at the surface leads to rising (sinking) air and cloudiness (clear skies). Convergence in the ITCZ is not continuous in time or space, because intermittently or locally, the trades are weak, notably in the regions known as the Doldrums in the eastern equatorial oceans. These regions were much feared by sailors in the days of sailing ships as vessels were becalmed there and ships could become stranded for long periods (see Figure 3.1). In the 1930s–1940s the ITCZ was termed the Intertropical Front, but although sometimes air mass contrasts may exist over land areas, the zone lacks the characteristics of mid-latitude frontal (baroclinic) zones. Subsequently, the term Equatorial Trough has been preferred, since it is descriptive rather than implying a mode of genesis. Nevertheless, the ITCZ is usually well developed across most of the Pacific and Atlantic Oceans. Satellite cloud studies show that convergence gives rise to cloud clusters, a few hundred kilometers wide, forming along the ITCZ (see cover plate).

The ITCZ migrates north and south, following with a 1–2 month lag the passage of the overhead sun. Figure 3.13 shows its latitudinal position in February and July in three regions of the globe. This analysis was based on 17 years of data on the presence of highly reflective clouds in satellite imagery. Such data represent the presence of deep cumulonimbus clouds. On average, the ITCZ is located at about 5 °N latitude because the thermal equator is displaced into the Northern Hemisphere due to the heated northern continents.

Figure 3.13
Latitude plots for
February and
July of the ITCZ
for three regions
based on highly
reflective cloud
data for
1971–1987
SP = surface
pressure;
OLR = outgoing
longwave
radiation; SST =
sea surface
temperature
(source: Waliser
and Gautier
1993).

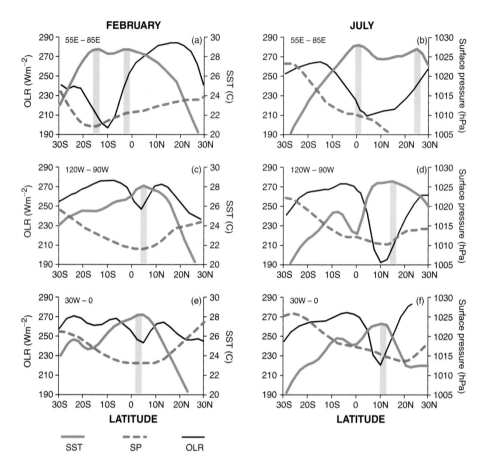

3.4 Storm frequency and tracks

Storms include *extratropical cyclones* in middle and high latitudes and *tropical cyclones* (hurricanes and typhoons). Cyclones are characterized by their diameter, central pressure value, and intensity. The airflow within them is anticlockwise in the Northern Hemisphere (cyclonic) and clockwise in the Southern Hemisphere as a result of the Coriolis effect being opposite in the two hemispheres. Extratropical cyclones are synoptic-scale features with a diameter of 1500–5000 km and a lifetime of 5–7 days. In the Northern Hemisphere, central pressures of mature systems average around 980–990 mb. Extratropical cyclones typically develop in the lee of mountain ranges (the Alberta and Colorado Rocky Mountains, and the European Alps), and along eastern coastlines of Asia, the Gulf Coast, and the Atlantic coast of the United States. Five storms per January move over the north-west Atlantic south of Newfoundland (Figure 3.14).

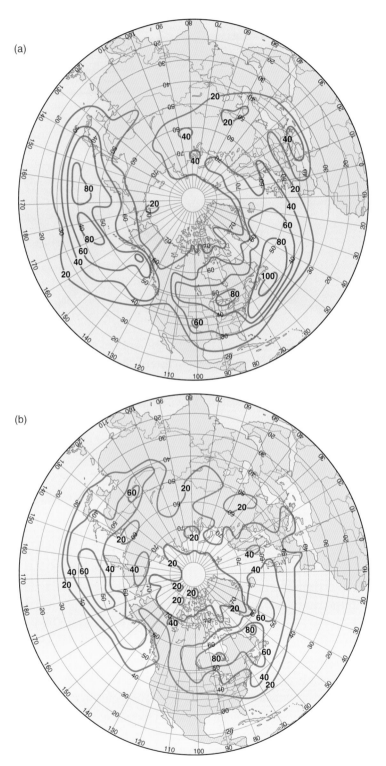

(a)

(b)

Figure 3.14
The frequency of cyclones in the
Northern Hemisphere in (a) January
and (b) July, for the 20 years 1958–
1977. Divide the values by 20 to
obtain annual frequencies (source:
Whittaker and Horn 1984).

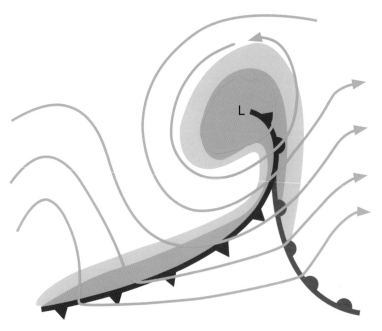

Figure 3.15
Illustration of an occluded front (source: Wikipedia, https://en.wikipedia.org/wiki/Surface_weather_analysis).

Cyclones tend to form either along air-mass frontal boundaries of warm and cold air, in the lee of mountain ranges, or beneath the eastern limb of an upper air trough where there is upper-level horizontal air divergence and associated upward motion that leads to falls in surface air pressure, hence deepening the surface low-pressure center. Such zones are more common over the eastern parts of the continents in winter. Cyclones decay when the upper-level divergence weakens, thereby lessening the upward motion. Surface inflow then fills in the cyclone.

Typically, several cyclones follow one another along one of the main frontal zones. The leading edge is a warm front, where warm air overrides colder air ahead of it. Behind is a cold front marking the leading edge of cold air in the rear of the system. Over the life cycle of the cyclone, the cold front catches up with the warm front and the warm sector is lifted off the ground. This process is termed *occlusion* and the combined front is known as an occluded front (see Figure 3.15). The airflow is typically organized into a warm conveyor belt that transfers moist air from the south in the warm sector (between the two fronts) northeastwards over the warm front (in the Northern Hemisphere). The cold conveyor belt is a rising (relative) airstream from the southeast, which is initially below the warm conveyor belt, but then emerges from below and extends northward.

Frontal cyclones, particularly over the oceans, tend to develop in sequence forming a "family," as illustrated schematically in Figure 3.16. The fronts in the leading system are generally occluded while the later ones have a more open wave form.

Figure 3.18 shows the distribution in each hemisphere of closed 1000–500 mb thickness lows, which are lows in the lower tropospheric layer. These are

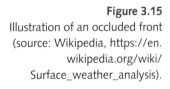

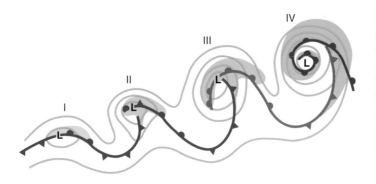

Figure 3.16
A sequence of marine frontal cyclones forming a "family": (I) incipient system; (II) cold front fracture; (III) bent-back warm front forming t-bone; (IV) warm core seclusion. Sea-level isobars and satellite cloud signatures (yellow) (source: Neiman and Shapiro 1993).

Box 3B.1 **Weather analysis**

The analysis of surface weather phenomena is carried out by the use of surface weather maps that utilize a plotting code to display the weather conditions at surface weather stations at a given time (usually every six hours: 00, 06, 12, and 18 hours UTC). Air temperature, dew point, cloud amounts and types of low, middle, and high cloud layers, wind speed and direction, MSL pressure and pressure change over the last three hours, visibility, present and past weather conditions are all plotted using numbers and symbols. Present weather is ranked according to its severity:

1 fine
2 weather in the past hour
3 blowing sand or snow
4 fog
5 drizzle
6 rain
7 snow
8 shower
9 thunderstorm.

An analyst uses the pressure data to draw isobars of equal pressure (usually at 4 or 5 mb intervals) and weather fronts. There are special symbols for different frontal types. Warm fronts are marked by semi-circles pointing in the direction of movement, cold fronts by triangles, and occluded fronts by alternating semi-circles and triangles (Figure 3.17). High-pressure centers are marked by H and low-pressure ones by L. This is known as a *synoptic weather map* since it shows synoptic-scale weather systems at a given time. They may be continental or hemispheric in scale. Figure 3.17 shows a sample weather map for the United States for 30 April 2012.

Box 3B.1 (continued)

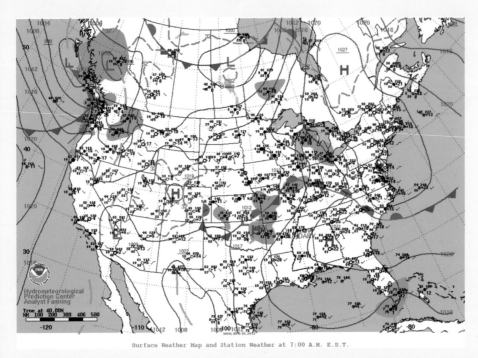

Surface Weather Map and Station Weather at 7:00 A.M. E.S.T.

Figure 3.17 Illustration of a surface weather analysis over the United States for 30 April 2012. Warm fronts are marked by red semi-circles and cold fronts by blue triangles. Both symbols are combined in the occluded front off the Pacific Northwest (source: daily weather maps: www.hpc.ncep.noaa.gov/dailywxmap/index.html).

Weather maps were first constructed after the weather occurred in the 1840s to study severe storms, when weather data were not immediately available. Following the invention of the telegraph, the Smithsonian Institution in Washington, DC, became the first organization to draw real-time surface weather analyses for the eastern and central United States in the late 1840s. Although frontal analysis began in Norway in 1918, fronts were not analyzed by the US Weather Bureau until 1942. Automated plotting of weather maps was undertaken in the United States in the 1970s. Now, with computers and geographic information systems, satellite and radar data can be overlaid on surface and upper air charts.

commonly cut-off lows in middle latitudes formed when the planetary wave structure becomes amplified and a high-pressure cell forms in high latitudes and a low-pressure in lower latitudes, separated from the main westerly wind belt. They tend to be slow moving systems.

Cyclones tend to travel northeastward in the Northern Hemisphere and southeastward in the Southern Hemisphere as a result of steering by the upper winds,

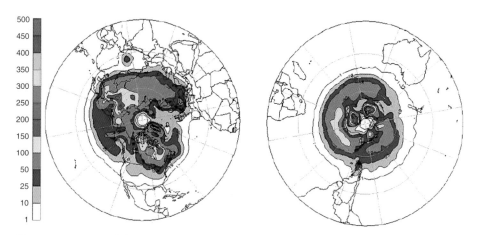

Figure 3.18 Maps of the occurrence of closed 1000–500 mb thickness lows, less than or equal to 5220 gpm, for the Northern Hemisphere (left) and Southern Hemisphere (right), 1951–2000 (source: Galambeau *et al.* 2004).

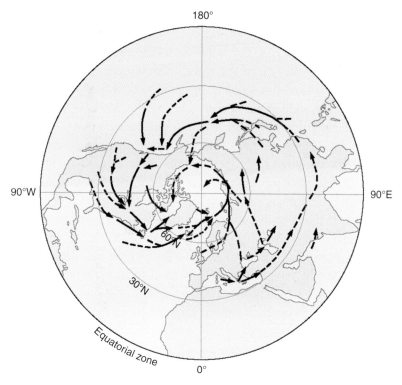

Figure 3.19 Cyclone tracks in the Northern Hemisphere in winter. The dashed lines show subsidiary tracks (source: Barry and Chorley 2010).

moving at around 1200 km day^{-1}. Figure 3.19 shows winter cyclone tracks in the Northern Hemisphere, which are seen to be closely related to the major frontal zones (Figure 3.11).

Tropical cyclones are usually much smaller than extratropical systems, with a radius of 200–1000 km. Pacific typhoons are typically twice the size of Atlantic hurricanes. They generally form in summer and autumn at least 5° latitude from the Equator (to acquire Coriolis force = $2\Omega V \sin \phi$, where Ω (omega) = Earth's angular velocity, V = wind speed, and ϕ (phi) = latitude angle), otherwise the air would flow directly into the low-pressure center. They also develop over areas where sea surface temperatures exceed 27 °C, through the rapid growth of a cyclone from a tropical wave or depression. The genesis appears to involve a warm core system with an ~ 100 km wide moist air column.

Worldwide, about 90 (50) tropical storms (cyclones) develop each year (see Section 8.2). At the tropical storm stage there are winds of 17–33 m s^{-1} (63–118 km hr^{-1}). The final stage involves the formation of an "eye," where winds are light and skies are largely clear. The eye has a diameter of 50–100 km and is surrounded by a cloud eye wall, where air ascends, and by a circular band of very high wind speeds. A category one hurricane on the Saffir–Simpson scale has sustained winds of 34–43 m s^{-1} (119–153 km hr^{-1}), and a category five hurricane has winds > 69 m s^{-1} (249 km hr^{-1}). Cloud and rain bands spiral inward toward the eye. At the top of the storm there is outward anticyclonic airflow maintaining the updraft. Occasionally, the rotation associated with tropical cyclones leads to the generation of tornadoes.

Information about conditions inside hurricanes is obtained via satellite remote sensing, weather radar, special aircraft reconnaissance flights, and dropsondes that aircraft release into the storm to gather pressure, temperature, humidity, and wind data as they descend, and transmit the data to hurricane centers.

During each season, tropical storms and cyclones are named in an alphabetical sequence of alternating female and male names. In 2010 there were 19 tropical storms and 12 hurricanes in the North Atlantic.

Tropical cyclones are primarily steered by the average flow in a deep atmospheric layer (~ 850–500 mb), modified by the increase of the Coriolis force toward the Poles (the so-called beta effect). When a steering flow is absent, the equatorial beta effect causes a cyclone in the Northern Hemisphere to move toward the northwest. The beta effect causes cyclones embedded in the tropical easterlies south of the subtropical ridge to move faster and slightly to the right of the steering flow. The speed of motion of tropical cyclones is generally 3–10 m s^{-1}. The storms quickly dissipate over cool ocean surfaces, which cut off the heat supply, and over land, which reduces the moisture source and increases surface friction, slowing the air motion. Sometimes a tropical cyclone may turn poleward, recurving around the western end of the subtropical high-pressure ridge, and redevelop in the westerlies as an intense mid-latitude cyclone.

3.5 Thunderstorms

Thunderstorms are a category of severe weather that involves lightning, strong winds, and generally heavy precipitation, although on occasion no precipitation will fall. Single-cell thunderstorms typically develop when strong surface heating, an unstable atmosphere, and moist updrafts build towering cumulus and cumulonimbus clouds that can extend through the troposphere up to a height of 15–20 km. They have a diameter of 20–25 km and a life span of an hour or less. The lower parts are made up of water droplets, while the tops are composed of ice crystals that may be blown by the upper winds into an anvil shape. Falling droplets sweep up others and grow by accretion into large raindrops. Hailstones may form on an ice nucleus of graupel (see Section 2.2d) and the cloud becomes electrified, producing lightning (see Section 2.2c). At the mature stage there are warm updrafts and downdrafts of cool air generated by falling precipitation.

There are three other types of thunderstorm – multicell clusters, multicell lines, and supercells. Multicell clusters comprise several storms in a cluster; each cell may last only 20–30 minutes, while the overall system travels generally eastward and may last for several hours. Mature cells are present near the storm center and dissipating cells on the downwind side. Ahead of, or at a cold front, a line of cells may develop, forming a squall line. They move with the front and bring heavy precipitation, hail, and strong winds. Supercell storms are extensive systems that move with the airflow toward the east or northeast. They are most common over the Great Plains of the United States from April to October. They are associated with a mesocyclone (a small-scale low-pressure cell 5–80 km in diameter) and a deep rotating updraft that in about 30 percent of cases leads to the formation of one or several tornadoes. They also occur in east-central Argentina and Uruguay, Bangladesh and parts of eastern India, South Africa, eastern Australia, and eastern China, as well as occasionally in the United Kingdom and Europe.

3.6 Mesoscale convective systems

Mesoscale convective systems (MCSs) are intermediate in scale between individual thunderstorm cells and synoptic cyclones and have a lifetime of about 6–12 hours. Cold, high cloud tops extend over an area of 50 000 km^2 or more. They develop as thunderstorms become clustered together, either in a linear pattern (a squall line) or as an amorphous elliptical cluster. They are common in spring and autumn in the central United States, eastern China, and South Africa and they also occur in the Tropics in India, West and Central Africa, and northern Australia. They bring severe weather with thunderstorms, heavy rain, hail, and sometimes tornadoes, especially in the US Great Plains. Typically, they develop in late afternoon and last into the night. In the northern areas of eastern China there are about 50 squall lines annually, with a peak in July in early evening. They are around 250 km long, oriented SW–NE and move eastward over about five hours.

A related system that occurs in the central United States in spring–summer is the derecho (Spanish for direct), which is a linear or cellular pattern of thunderstorms with strong winds above 25 m s^{-1}. The area extending northeast from the southern Plains to the middle Mississippi valley experiences one such event per year. Twenty percent of cases occur each in May, June, and July. They are associated with thunderstorms that develop a forward bow shape. The strong winds are a result of convective downdrafts or microbursts. Serial derechos are embedded in an extensive squall line hundreds of kilometers wide, while progressive derechos are relatively narrow but travel great distances.

SUMMARY

Atmospheric pressure is due to the weight of the overlying air column. Globally, there is an equatorial belt of low pressure, subtropical highs, and subpolar lows. MSL pressure is ~1013 mb. High-pressure centers average about 1040 mb and mid-latitude lows about 980 mb. Winds blow anticlockwise around lows (highs) in the Northern (Southern) Hemisphere. Geostrophic flow represents a balance between the pressure gradient force and the Coriolis force in the opposite direction due to the Earth's rotation. Wind systems transport moisture around the globe. Notably, there are well-defined systems of atmospheric rivers in the lower troposphere.

Air masses are large bodies of air with considerable horizontal homogeneity of temperature and moisture conditions. They are defined by temperature (polar, tropical) and continental/maritime origin. Air mass boundaries define frontal (baroclinic) zones. The polar frontal zone is a major feature of middle latitudes in both hemispheres. Frontal low-pressure systems develop along it at 3–7-day intervals and move generally eastward between latitudes 40° and 60°. Tropical cyclones develop from wave depressions in the tropical easterlies over waters with a surface temperature ≥27 °C. They are most frequent in summer–autumn in the tropical western oceans. In the western North Pacific they are known as typhoons, and in the western North Atlantic as hurricanes. They bring heavy rains, very strong winds, and storm surges, causing widespread damage and loss of life. Cloud bands spiral around an inner eye (50–100 km across) where winds are light and skies may be clear.

Thunderstorms develop with strong surface heating, an unstable atmosphere, and moisture, which leads to the build-up of towering cumulus and cumulonimbus clouds. They give rise to heavy precipitation, lightning, and strong winds. Thunderstorms may be single-cell, multicell clusters, linear squall lines, and super-cells. Mesoscale convective systems are thunderstorm clusters with a life span of 6–12 hours. They are associated with severe weather, including tornadoes, and particularly affect the US Great Plains in spring and summer.

QUESTIONS

1   The geostrophic wind blows at right angles to the isobars. True or false?
2   The pressure gradient is steeper in low pressure than high pressure. True or false?
3   Compare the wind speed variation with height in the westerlies and tropical easterlies.
4   What factors lead to air mass contrasts?
5   The polar front forms between continental Polar and maritime Polar air. True or false?
6   Compare the life cycle of a frontal depression and a mesoscale convective system.
7   What are the characteristics of the ITCZ?
8   Describe the weather map shown in Figure 3.17.

4 Local and microclimates

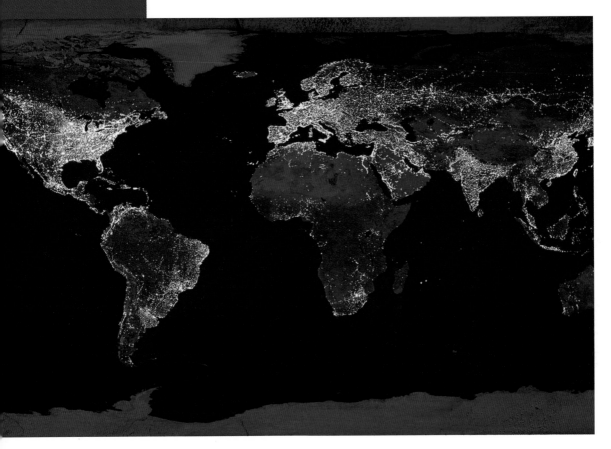

Global night lights.

S O FAR we have talked about large-scale climatic conditions, but climate also varies considerably on local (1–10 km) and micro (1–10 m) scales. Examples of the former are seacoasts, mountains, forests, lakes, and urban areas. Examples of the latter are plant and forest canopies and urban canyons. It is important to recognize the effects of these small scales on climate because observations are made at a point and the representativeness of the location determines our ability to scale up the information to a wider region. In addition, humans experience microclimates in their everyday lives.

4.1 Local climate

The controls of *local climate* are slope orientation and angle, large-scale shelter and local winds due to topography, and surface energy and water budgets. Topographic effects give rise to what are called *topoclimates* that have a horizontal scale of about 1–10 km. Figure 4.1 illustrates topoclimates associated with a lake, forest, and urban area.

The direction a slope faces has a large impact on sunshine duration and shadow, while slope orientation and angle affect the amount of solar radiation that is intercepted. These same factors are critical in the installation of solar panels on a roof. In the Northern (Southern) Hemisphere, radiation totals are largest on south (north)-facing slopes. For example, measurements in Vienna (48° N) on 20° slopes show that direct beam radiation in July ranges from 352 MJ m^{-2} on east/west slopes to 311 MJ m^{-2} on north-facing and 389 MJ m^{-2} on south-facing slopes, compared with 373 MJ m^{-2} on level ground. In December, the corresponding values are: 31, 0, 67, and 34 MJ m^{-2}, respectively. The percentage differences are much larger in winter.

Large-scale shelter refers to the air drainage into a valley or basin on a scale of ~ 10 km radius. At night, cold air, produced by *radiative cooling* of the ground on the upper slopes, drains down slopes into basins, because it is denser than the surrounding air. This drainage forms a cold air lake that by the end of the night may be several hundred meters deep. The inversion capping it is located at about 0.20–0.25 of the relative relief (the height range between the valley bottom and ridge top). In the morning, the slopes become heated and the warmed air mixes upward into the base of the inversion, eroding it. As a result, the top of the inversion lowers until, ultimately, the basin is full of thoroughly mixed air.

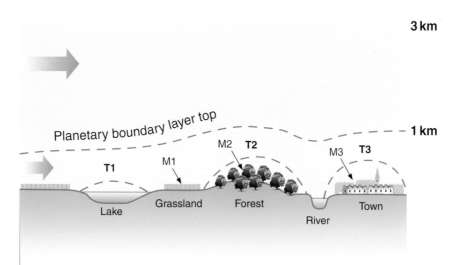

3 km

Figure 4.1
Schematic illustration of topoclimates (T1, T2, and T3) and microclimates (M1, M2, and M3).

The middle zone of a slope is typically 1–3 °C warmer at night than the hilltop or basin. Appalachian farmers first observed this feature in the mid nineteenth century. In the Middle Mountains (Mittelgebirge) of Europe, such as the Vosges and Jura, where the relative relief is about 500 m, the warm zone (so-called thermal belt) is located 100–400 m above the valley floor. In the Alps the thermal belt is about 350 m above the valley floor in summer and 700 m above the floor in winter.

Local winds are a result of horizontal and vertical gradients of temperature and therefore density of the air. At the seacoast or at the coast of large lakes, sea and land breezes develop whenever the large-scale pressure gradient is weak and there are land–water temperature contrasts. By day, the land is heated, setting up a local low-pressure area. A breeze off the water develops toward the low pressure and the air over the low pressure rises a kilometer or so and then flows out over the water as a return current, making a local vertical circulation cell (see Figure 4.2a). The sea or lake breeze may form a local front of cool air that progresses inland, sometimes up to 50–100 km, during the day. A line of cumulus cloud may form along the cold air front due to the moisture evaporated off the water. During the day, the sea or lake breeze turns to the right (in the Northern Hemisphere), as a result of the Coriolis effect, and eventually blows parallel to the coast. The breeze dies down in the evening and, as the land cools, a land breeze blows toward the warmer water. This also sets up a reverse circulation cell, analogous to the daytime one (Figure 4.2b). The sea breeze has speeds of about 5–10 m s^{-1}, and the land breeze about half of that due to the weaker temperature gradient at night.

Seacoasts also give rise to zones of convergence or divergence, depending on the wind direction. Where the wind is blowing with the land on the left (right) side of the airflow in the Northern (Southern) Hemisphere, the change in friction across the coast results in the flow diverging (converging). This is exemplified along the Caribbean coast of Venezuela when there is easterly flow in spring and summer, giving rise to coastal divergence. Convergence leads to uplift, forming clouds and sometimes leading to precipitation, while divergence leads to subsiding air and clear skies. Convergence (divergence) also occurs where the wind is perpendicular to the coast and onshore (offshore).

On mountain slopes there are corresponding downslope (katabatic) winds at night, as cold dense air drains downslope, and upslope (anabatic) winds by day due to slope heating. Both systems are about 100 meters deep and have speeds of 2–5 m s^{-1}. They develop when the general pressure gradient is weak and skies are clear. Mountain valleys set up wind systems along the valley axis when large-scale pressure gradients are weak and there is sufficient daytime heating/nighttime cooling. On summer days the limited air volume in the valley is heated 2–3 times more than the larger air volume over the adjacent lowland. This sets up a

(a)

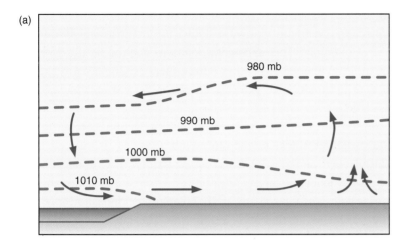

980 mb

990 mb

1000 mb

1010 mb

(b)

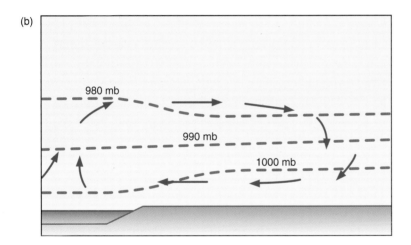

980 mb

990 mb

1000 mb

Figure 4.2
Illustration of the daytime sea breeze (a) and nighttime land breeze (b) at a coast (source: Barry and Chorley 2010).

density difference and so air flows up-valley in a valley wind by day. It has speeds of 3–5 m s^{-1} and is accompanied by a return current above the valley, forming a local circulation cell. Valley winds are best developed where a valley is incised into a plateau. At night, a cold mountain wind blows down valley, again with a return current above it at about ridge-top level. The slope winds form a part of the whole mountain–valley wind system. The wind systems transition from one mode to the other over about an hour's time, after sunrise and sunset.

On the scale of an entire mountain range there are also regional wind systems. By day, if the synoptic pressure gradient is weak, there is a plains-to-mountains wind, and at night there is a reverse circulation from the mountains to the plains. These systems are about as deep as the terrain. Their existence has been demonstrated for the Alps and the Colorado Rocky Mountains, but they have been little studied.

Mountain ranges have major impacts on all climatic elements.

When air is forced to rise up the windward slope it is moving to a lower-density environment, so it expands. This expansion leads to cooling as the work of expansion uses internal energy. A process where there is temperature change without addition or subtraction of external energy is known as adiabatic. (A diabatic process is one where there is an external heat source or sink, as in the cases of solar and terrestrial radiation.) As discussed in Section 3.1b, air that is unsaturated cools adiabatically at a fixed rate of 9.8 °C km^{-1}, the dry adiabatic lapse rate (DALR). Hence, air rising 2000 m cools by almost 20 °C. If the air is still unsaturated when it reaches the mountaintop it can warm at the same rate as it descends on the lee side. However, rising air that reaches condensation level will release latent heat as it rises further and cloud forms. This latent heat slows the cooling by an amount dependent on the moisture content of the air. This rate is known as the saturated adiabatic lapse rate (SALR), which is typically about 5–6 °C km^{-1}.

Where air ascends and cools at the SALR, and then descends and warms at the DALR, there will be a net rise in air temperature in the lee. This warming also leads to a decrease in relative humidity. This is one type of *foehn* (in the Alps) or *chinook* (in the Rocky Mountains) mechanism. Chinook is an indigenous word meaning "snow eater," as it causes the rapid sublimation of snow cover in the lee of the Rocky Mountains. The same effect is achieved when air is forced to descend from summit level to the lee side as a result of blocking of the airflow on the windward side by a stable layer located at the mountaintop. Across the Alps there may be south foehn from the Mediterranean or north foehn from Bavaria, depending on the pressure pattern. The Santa Ana (foehn) wind in southern California, blowing from the high elevations of the Great Basin and Mojave Desert to the east, brings a high risk of fires in the chaparral (evergreen shrub) vegetation in autumn due to the very dry, hot air.

When the air that is crossing the range is very cold, the air descending the lee slope may warm adiabatically but still arrive at the foot of the mountain cooler than the air it displaced. This occurs in winter across the Dinaric Alps in the vicinity of Trieste, Italy, where the cold wind from southeastern Europe is known as a bora. It is dry, strong, and gusty, blowing out over the Adriatic Sea.

The surface energy and water budgets are determined by the surface characteristics – the type of vegetation or land use, the albedo, and soil type. Swamps and wetlands have a high heat capacity and so warm up and cool down slowly. Most of the energy exchange is via latent heat. Arid surfaces, on the other hand, heat up and cool down rapidly and almost all of the heat exchange is via sensible heat. Albedos range from about 0.12 for wetlands to 0.35 for desert surfaces. The role of soil types is discussed in Section 4.5.

4.2 Forest climate

Forests cover vast areas of the globe in the tropical rainforests of the Amazon, the Congo, and Indonesia, and the boreal forests of Canada and the Russian Federation. In total, forests cover about 33 percent of the land surface. In South America and in the Russian Federation this figure rises to 49 percent. Forests have major effects on wind and radiation conditions compared with their exterior surroundings. Trees in tropical rainforest can grow to 65–70 m in height, although the average canopy is usually about 30 m high. The light intensity on the floor of the rainforests is about 0.5 percent of that at the canopy, although light flecks may increase this threefold. In Rondônia, Brazil, the net radiation at the surface was only 4 percent of that above the canopy. In comparison, in mid-latitudes the fraction of light reaching the forest floor is 10–25 percent for spruce and fir, 20–40 percent for pine, and 50–75 percent for birch–beech forest.

With respect to the forest edge, light conditions change almost immediately, while wind speed, temperature, and vapor pressure adjust over some 40–50 m into the forest when the wind is directed into it. Gaps and clearings in forest cover give rise to higher solar radiation receipts, air temperatures and evaporation up to an area of about 600 m^2, beyond which there are no changes.

In summer the dense shade of forest canopies lowers the daily maximum temperature by 1–2 °C compared with outside. Conversely, nighttime lows are slightly higher inside the forest. Compared with the top of the canopy, the daily temperature range in the trunk space of rainforest in Nigeria is about half (4 °C compared with 8–9 °C). Daytime temperatures near the floor of tropical rainforests are typically some 7–10 °C below those at the canopy.

Wind speeds drop dramatically inside a forest. Measurements in European forests show that wind speeds are reduced to 10 percent of those outside within 100 m of the forest edge. As a result of the light winds there is an increased frost risk in forest clearings. Forests also have a filtering effect on pollutants in the air. Measurements of particles downwind of a 1 km wide forest in Germany were half those on the upwind side, for example.

Interception of rainfall by the tree canopy plays an important role in forest water balance. Measurements in temperate pine forest indicate that about 30 percent of annual rainfall is intercepted, while in tropical rainforests about 13 percent is intercepted due to the higher rain intensities in the Tropics. The intercepted water evaporates from the canopy, or reaches the ground by through-flow. Hence, in the latter case the interception is not a total loss to the forest water balance.

Snow accumulation in a forest depends on whether it is evergreen or leafless. Evergreen canopies retain much of the snowfall, from where some of it may sublimate (turn directly into vapor) into the atmosphere. As a result, beneath canopies in the boreal forest there is about 65 percent of the snow water equivalent (SWE) measured in open areas. Sublimation requires the supply of the latent heats of both fusion (solid to liquid) and vaporization (liquid to vapor). This amounts to nearly eight times the energy for melt, and hence sublimation is less common. It is mainly significant in dry, sunny, windy climates and from blowing snow. Sublimation accounts on average for the loss of about 15–25 percent of the seasonal snowpack.

4.3 Lake climate

Worldwide there are about 1.5 million lakes exceeding 10 hectares (25 acres) in area, which makes up ~2.1 percent of the land surface. Satellite estimates based on the Advanced Microwave Scanning Radiometer (AMSR)-E give a mean total water area of ~3.0 percent and an annual maximum of ~3.8 percent of the land area. Large water bodies – like the Great Lakes, Great Slave Lake, and Great Bear Lake of North America – exert important local effects on the climate over and around them. They are considerably cooler than the adjacent land in summer and warmer in winter, before they freeze up. Open water at this time of year leads to high evaporation rates into cold, dry cP air flowing over them, forming convective cloud. Frictional convergence of the airflow at the lake shoreline triggers convective precipitation and heavy snowfalls in early winter. This process is known as "lake effect." Syracuse, NY, adjacent to Lake Ontario, receives almost 300 cm of snow annually. Locations in the Upper Peninsula of Michigan exceed 700 cm per year. Lake effect snow belts typically extend 30–40 km inland (Figure 4.3). As the lakes cool down and begin to freeze, the lake effect snowfalls diminish in intensity and extent.

The lake effect obviously depends on the fetch, or distance that the air travels across the lake. For example, the flooding of the dry salt pans of Lake Eyre (8000 km^2), Australia, in 1949–1950 was found not to increase the local rainfall, according to a team sent to investigate.

The effect of the Great Lakes on air temperatures is estimated to be about $+2.8\,°C$ in January and $-1.7\,°C$ in July compared with their surroundings. Across Lake Michigan differences are similar. For the five coldest Januaries between 1871 and 1930 the mean temperature at Milwaukee (43° N) on the west side was $-12.4\,°C$, and $-8.4\,°C$ at Grand Haven on the east side. Stations around Lake Baikal (31 000 km^2) in Russia show larger effects, according to P. Crowe. In December, before the lake freezes the warming is around 9 °C; in July the cooling is 7 °C. In equatorial East Africa, Kampala (1312 m) on Lake Victoria (67 000 km^2) is 5 °C cooler during December–January and 3 °C in June–August than the town of Mbale (1220 m), 110 km to the northeast.

Regions with extensive water bodies have effects on regional climate. Simulations with a global climate model by G. Bonan show that the inclusion of

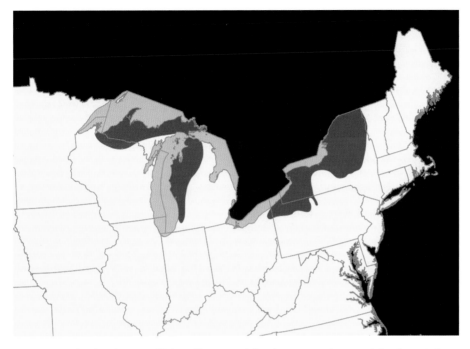

Figure 4.3 The distribution of lake-effect snowfalls, shown in red, around the Great Lakes (source: Wikipedia, http://en.wikipedia.org/wiki/Lake-effect_snow).

lakes and swamps in the model for July led to temperatures that were 2–3 °C lower, increased latent heat flux by 10–45 W m^{-2}, and decreased sensible heat flux by 5–30 W m^{-2} compared with the simulation without these water bodies. The changes were statistically significant in the lake region of northern Canada, the Great Lakes region of North America, the swamp and marsh region of the Siberian lowlands, and the lake region of East Africa.

4.4 Urban climate

In 2008 the urban population of the world accounted for half of the global population and there were 19 cities with ten million or more inhabitants. Thus, the climate of urban areas is of vital importance for humanity. The earliest study of city effects on local climate was carried out by Luke Howard in England. In 1818 he published a book on London's climate based on data collected at a number of weather stations in and around the city. London was also the subject of research by T. Chandler in the 1960s, using cars fitted with measuring instruments to make transects of the city at night.

Cities have major effects on air quality, as well as modifying the incoming solar radiation, air temperature, moisture conditions, wind regime, and visibility. The main reasons for these modifications are the changes in surface characteristics (brick, concrete, asphalt, and glass), the presence of tall buildings and urban canyons, the combustion of fossil fuels for heating homes and offices in winter, for transportation, and industry, and the production and release of pollutants.

(a) Pollution

Pollutants in urban areas consist of particulates (compounds of carbon, lead, aluminum, and silica) and gases (sulfur dioxide, carbon monoxide, ozone, hydrocarbons, and nitrogen oxides). Particle concentrations in industrial cities may reach 50–100 μg m^{-3}, compared with rural levels of 20–30 μg m^{-3}. The main effects of particulates are to reduce incoming solar radiation (see Section 2.1c) and visibility and to serve as cloud condensation nuclei. This last process was first investigated by Helmut Landsberg (see Box 4B.1). Ozone concentrations of 0.35 ppm can also lower global shortwave radiation by 15 percent. Pollution may be trapped at night by low-level temperature inversions, leading to the formation of a pollution dome over the city.

(b) Urban heat island

The heat budget of cities is modified in several ways. First, there is heat released by domestic combustion; in mid-latitude cities in winter this is comparable with the

Box 4B.1 **Helmut Landsberg**

Helmut Landsberg was born in Frankfurt, Germany, in 1906. He studied geophysics, and in 1934 moved to Pennsylvania State University where he taught geophysics and meteorology. He also offered the first graduate seminar on bioclimatology in the United States. He initiated the statistical analysis of climatic data and used these methods to provide climatic summaries for military operations during World War II. In 1949 he was appointed to President Truman's Air Pollution Committee, where he helped shape air pollution regulations. From 1954 to 1966 he was Director of the Environmental Data Service at the US Weather Bureau. In 1967 he became the first Director of the Graduate Program in Meteorology at the University of Maryland, where he remained until his death in 1985. He is best remembered for serving as Editor-in-Chief of 15 volumes of the *World Survey of Climatology* from 1964, but he also published *Urban Climate* (last edition in 1981).

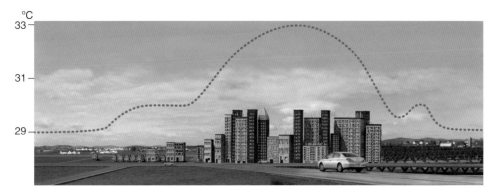

Figure 4.4 Schematic cross-section of a city showing the urban heat island. The temperature scale refers to the red dashed line (source: NASA).

incoming solar radiation. Second, there is heat storage in city buildings and road surfaces due to the absorption of solar radiation during the day. The albedo of city surfaces is of the order of 0.12–0.14, about half that of grass. This heat is released at night by infrared radiation to the lower atmosphere and by sensible heat transfer to the air, contributing to an *urban heat island*, relative to the surrounding rural areas. Figure 4.4 illustrates schematically how the temperature is several degrees higher than in the rural areas. In North America, urban–rural temperature differences are 8 °C for cities of 100 000 people and 12 °C for cities of one million. Differences are less in European cities, perhaps due to the generally lower buildings. Differences are greatest for nighttime minimum temperatures.

Table 4.1 Daily maximum urban heat island intensity versus latitude (from Wienert and Kuttler 2005)

Latitude zone	Urban heat island (°C)
0–20°	4.0
20–40°	5.0
40–60°	6.1

The intensity of the urban heat island shows a relationship with latitude. Table 4.1 shows the results of a study by Wienert and Kuttler. The effect is mainly related to anthropogenic heat production. However, the variance of temperature that is explained by latitude is only 6 percent.

The factors that contribute to heat island intensity have been quantified for a mid-latitude city in summertime using a mesoscale atmospheric model by Ryu and Baik. The three main causative factors are: anthropogenic heat, impervious surfaces, and three-dimensional (3D) urban geometry. The 3D geometry factor is subdivided into: additional heat stored in vertical walls, radiation trapping, and wind speed reduction. In the daytime, the impervious surfaces contribute most to the heat island intensity. The anthropogenic heat contributes positively, whereas the 3D urban geometry contributes negatively. In the nighttime, the anthropogenic heat contributes most to the intensity, followed by the impervious-surfaces factor. The 3D urban geometry contributes positively to the nighttime intensity, primarily via the additional heat stored in vertical walls.

In comparing urban and rural temperatures, care must be taken in the selection of station sites with respect to topography, the location of parks, and water bodies, to ensure that representative locations are selected. The definition of urban–rural differences is problematic as there are many different types of urban environment. For this reason, a series of 17 *local climatic zones* have been defined by Stevens and Oke. Ten are built types and seven are land cover types. The built types range from compact high-rise to sparsely built, with a separate class for industry. The land cover classes range from dense trees to bare surfaces and water. Each class is determined from the following variables: sky view factor, aspect ratio, the fraction of building surface, the fractions of impervious and pervious surfaces, the height of the roughness elements, and the terrain roughness class. This approach offers a new methodology for urban climate studies.

In some mid- and high-latitude cities there appears to be a "cold island" or "cold oasis" effect of 1–3 °C in summer. This is attributed to shading in urban canyons. In Phoenix, Arizona, the surrounding dry semidesert warms at a greater rate than the urban complex during daytime, setting up a (relatively) cold oasis.

Heat islands also occur in large tropical cities, but there are differences from mid-latitude ones as a result of generally low building height and different construction materials. Thermal inertia tends to delay the maximum temperature effects from nighttime to around sunrise.

Cities tend to have lower relative humidity than rural areas as rainwater runs off into sewers. Hence, there are few water sources for evaporation. This dryness enhances the city heating effect.

In recent years there has been growing awareness of the negative consequences of urban heat islands on energy bills (due to increased usage of air conditioning) and on human health. Efforts are being made to plant grass on rooftops, for example, to increase the albedo and lessen the absorption of solar radiation.

(c) Moisture effects

Cities also modify the local moisture budget as a result of the replacement of a vegetated land surface and water bodies by brick, concrete, and asphalt, leading to rapid drainage of rainwater and the great reduction of evapotranspiration into the air. As a consequence, relative humidity in the city is less than in its rural surroundings; at night the difference can be as much as 30 percent.

Urban influences on precipitation are harder to discern as a result of the large spatial and temporal variability of rainfall totals and the paucity of city rain gauges. However, there is evidence that the heating of mid-latitude cities in Europe and

North America leads to a small increase of 6–7 percent in the number of annual rain days compared with the surrounding areas. The maximum effects are typically observed a short distance downwind of the built-up area. Urban areas appear to trigger convection in marginal weather situations in summer. Summer thunderstorm rainfall over London has also been shown to have increased. Cities in the Midwest and southeastern United States also experienced 30 percent more rainfall than their surroundings on summer days with mT air.

4.5 Microclimate

Microclimate is often defined as the climate within the plant canopy, which may reach a height of 50 m in tropical rainforest, but only a few centimeters in Arctic tundra (see Figure 4.1). An important factor in microclimate is the surface boundary layer (~ 10–50 m thick) where wind speeds increase rapidly with height. This vertical wind shear, together with the surface roughness, gives rise to small-scale, chaotic (turbulent) air motion. Turbulence may be simply mechanical, due to the surface roughness and friction effects on the wind, or it can be convective, due to heated up-currents of air. This motion is important for the dispersion of pollutants, as well as for the vertical transports of sensible and latent heat.

The primary characteristic of microclimates is the large diurnal range of temperature at the surface and in the upper soil layer compared with that in the air at the standard 2 m height in a weather instrument shelter (Stevenson screen). The latter is a white-painted louvered wooden box standing about 1.5–2 m high, which contains dry- and wet-bulb thermometers and maximum and minimum thermometers. The louvers permit airflow and exclude solar radiation. The ground surface absorbs solar radiation and so it responds rapidly to this heating. The heat is communicated by conduction to an air layer a few millimeters deep, but this is then transferred slowly to the air above by convection, meaning that the air temperature lags the surface temperature by 1–2 hours. The difference between the surface and the screen air temperature may reach 10–15 °C. Likewise, the surface cools down more rapidly at night due to outgoing infrared radiation emitted by the ground.

If the air is moist, dew may be deposited on the surface. If the air is cold, so that the dew-point temperature is below freezing, hoar frost may be deposited on the surface (ground frost), although the air temperature may not reach freezing.

Soil type and soil moisture content play a considerable role in microclimate. Sandy soils have a low heat capacity and so heat up and cool down rapidly over the daily cycle. Clay soils contain more moisture and so have a high heat capacity, warming up and cooling down slowly. The penetration of the diurnal temperature wave into a sandy soil is also greater than in a clay soil. For example, on clear summer days at Sapporo, Japan, the diurnal temperature range at the surface is 40 °C on sand and 21 °C on clay; at 10 cm depth the corresponding values are 7 °C and 4 °C, respectively.

Surface skin temperature depends strongly on surface characteristics. The world-record surface skin temperature, as determined by the Moderate-resolution Imaging Spectrometer (MODIS), averaged 68 °C over 2004–2009 and reached an extreme of 71 °C on a black lava surface in the Lut Desert of southeast Iran, according to Mildrexle *et al.*

The presence or absence of winter snow cover has a major effect on air temperature. In the US Great Plains winter air temperatures are 5–10 °C lower in the presence of snow cover than without it. Deeper snow (> 30 cm) reduces air temperatures by about 1 °C compared with shallow snow (< 15 cm) as a result of the decreased upward sensible heat flux from the ground.

A particular aspect of microclimate concerns the response of the human body to heat and cold. This topic is treated in the next section.

4.6 Human bioclimatology

The human body can, with clothing, tolerate air temperatures ranging from about –60 °C to 60 °C. However, a clothed individual at rest feels most comfortable with an environmental temperature of about 20–25 °C. The body is a homeostatic system that can regulate its temperature so that the internal core remains close to 37 °C. Thus, under hypothermal (cold) stress the body shivers when the skin temperature drops below 20 °C, releasing heat by muscular action. Also, blood vessels contract (vasoconstriction) to decrease the flow of heat to the skin. When there is hyperthermal (heat) stress, the body responds by sweating when the skin temperature exceeds 36 °C to cause evaporational cooling. In addition, blood vessels near the skin dilate (vasodilation). The transition from vasodilation to vasoconstriction occurs at a neutral point at an "equivalent" temperature of 29 °C. This equivalent (or operative) temperature is the average of the mean radiant and ambient air temperatures weighted by their respective heat transfer coefficients. The flow of heat from the core to the skin decreases rapidly as the temperature difference between the core and the skin exceeds about 5 °C, but increases rapidly with vasodilation.

The energy balance of a human can be written as:

$$Rn + M = H + LE \pm \Delta S,$$

where M = heat production by metabolic processes and ΔS = net heat storage by the body. The other terms are as defined in Section 2.1g. The body gains heat through solar radiation and sensible heat from warm air, and loses it through thermal radiation, convection of sensible heat (H), evaporation from the lungs and sweating (LE). The fluxes are averaged over a standardized body surface area of 1.7 m^2.

The body at rest has a basal metabolic rate (BMR), which is the rate at which energy is expended to maintain life. This rate is about 58 W m^{-2}. About 70 percent of a human's total energy expenditure is due to the life processes within the organs of the body. About 20 percent of one's energy expenditure comes from physical activity. The metabolic rate increases from about 60 W m^{-2} when seated to 120 W m^{-2} when walking at 3.2 km h^{-1}, to 270 W m^{-2} when walking up a 15 percent grade, according to T. Oke. At rest, the chest and abdomen generate about 56 percent of body heat and the brain 16 percent. When active, the skin and muscles generate about 73 percent, the chest and abdomen 22 percent and the brain 3 percent.

At low temperatures wind has a major effect on skin temperature. A new *wind chill index* was recently developed that realistically takes account of clothing and the exposed face of a person walking (Figure 4.5). It also uses the wind speed at face height, 1.5 m. For example, a temperature of 0 °C with a 20 km h^{-1} (5.5 m s^{-1}) wind feels like − 5 °C.

Frostbite in the extremities (fingers, toes, nose, ears, cheeks) is a risk at high wind-chill values. Here, the skin and sometimes deep tissues freeze when the skin temperature reaches 0 °C. When the body cannot maintain its core temperature, *hypothermia* results. The subject becomes unconscious at a core temperature of about 30 °C and the heart usually stops at about 26 °C. Hypothermia can occur at above freezing temperatures, especially in cold, wet conditions.

Clothing insulation is measured in units of "clo," where one clo maintains comfort for a seated person indefinitely at an ambient temperature of 21 °C, with a wind of 10 cm s^{-1} and relative humidity below 50 percent. Polar clothing

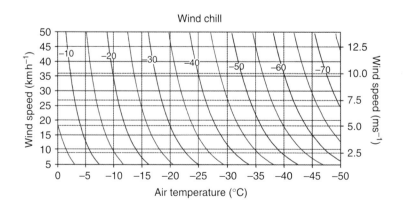

Figure 4.5
The index of wind chills equivalent temperature versus air temperature and wind speed (source: based on Osczevski and Bluestein 2005).

has a clo value of 3 or more. In the Alps at 1300 m, moderate activity requires 2.5 clo in summer and 4.5 clo in winter. Vigorous activity reduces these requirements to 2.0 and 2.5, respectively.

Heat indices are used when it is hot and humid in order to assess the perceived (apparent) temperature. The relative humidity is scaled to the air temperature to determine a perceived temperature. At 27 °C, the heat index will agree with the actual temperature if the relative humidity is 45 percent, but at 43 °C, relative-humidity readings above only 17 percent will make the heat index higher than 43 °C. The heat index table is at www.nws.noaa.gov/os/heat.heatindex.png.

Exposure to direct sunlight can increase the index by up to 10 °C. The NOAA indicates the following health hazards for given temperature ranges:

- 27–32 °C: caution, fatigue is possible.
- 32–41 °C: extreme caution, heat cramps and heat exhaustion are possible.
- 42–54 °C: danger, heat cramps and heat exhaustion likely; heat stroke probable with continued activity.
- > 54 °C: extreme danger; heat stroke imminent.

In the United States, statistics relating to deaths from heat and cold are contradictory, according to Dixon *et al.* One source reports twice as many deaths per year related to cold exposure during 1979–2002 as succumbed to extreme heat. Another source gives an average of 665 deaths per year from excessive heat and 748 per year from excessive cold for 1979–1999. Cases of hypothermia were most common during December, January, February (DJF) in Alaska, the northeast and upper Midwest, and hyperthermia (heat exhaustion, heat stroke, and sun stroke) in southern states in July. The extreme heat over the central United States in July 2010 is illustrated at http://disc.sci.gsfc.nasa.gov/gesNews/ steamy_heat_on_plains#news_item_image

High altitude imposes other stresses on the body. The decrease in air pressure with altitude affects the amount of oxygen that can be inspired, and this can lead to *hypoxia* – a lack of oxygen. About half the population may experience mild symptoms of mountain sickness (headache, nausea, shortness of breath) at 3000 m (700 mb) and most will experience these at 5500 m (500 mb). This latter elevation is about the limit of permanent human habitation in the Andes and Tibet. Tibetan populations seem to have acquired genetic adaptations to the hypoxic environment. Severe forms of mountain sickness can involve pulmonary edema (fluid in the lungs), which can be fatal if the person is not given oxygen or moved to lower altitudes. Nevertheless, some climbers have reached Himalayan peaks above 8000 m (370 mb) without oxygen. Hypoxic effects are often combined with hypothermia and dehydration due to the low temperatures and low vapor content of the air, adding to the hazards of high altitude.

SUMMARY

Local or topoclimates extend over a few kilometers horizontally, whereas micro-climates exist within a canopy. Major types of local climate are forests, lake and seacoasts, mountains, and urban areas. An important factor in topoclimate is slope orientation and angle, which affects incoming solar radiation amounts and therefore temperatures. Land–water boundaries give rise to diurnal sea (or lake) and land breezes, with an overlying return current. Valley slopes have diurnal anabatic and katabatic winds, while along the valley axis are valley and mountain winds, with upper-level return flows, as a result of heating contrasts between the plains and valley. Large lakes give rise to lake effect snows on the lee shores in autumn and winter before the lake freezes. There are also temperature contrasts between the upwind and downwind shores. Forests have a mitigating effect on temperatures, wind speeds, and aerosol concentrations compared with outside the trees. They have complex effects on the energy and water balance and on snow retention.

Urban climates are determined by the surface materials and building height, as well as by the domestic heat emitted in winter. The atmosphere is modified by pollutants, which affect air quality, visibility, and solar radiation receipts. Cities are typically warmer than their rural surroundings (urban heat island), especially at nighttime in winter.

Human bioclimatology addresses the body's response to heat and cold stress. Threshold values for hyperthermia and the risk of heat stroke depend on temper-ature and atmospheric humidity, while hypothermia and frostbite risk depend on temperature, wind speed, and clothing worn. High altitude (>3 km) increases the risk of mountain sickness as a result of hypoxia. The limit of human habitation is ~5500 m.

QUESTIONS

1   Compare the main factors that determine local and microclimates.
2   An anabatic wind blows which of the following directions:
   (a) upvalley;
   (b) downslope;
   (c) onshore;
   (d) upslope;
   (e) downvalley?

3   Discuss the seasonal variation in lake effects on climate.
4   Compare the climatic conditions inside and outside a forest.
5   Consider the major differences between the climate of a large city and its rural surroundings.
6   Explain when and why vasodilation and vasoconstriction occur.
7   Altitude sickness is caused by which of the following:
    (a) hypothermia;
    (b) low air pressure;
    (c) hypoxia;
    (d) vasodilation?
8   The sea breeze turns to blow parallel to the shore during the day. True or false?
9   Compare the *foehn* and the *bora* and their effects.
10 Contrast human responses to heat stress and cold stress, both physiological and cultural.

The general circulation

5

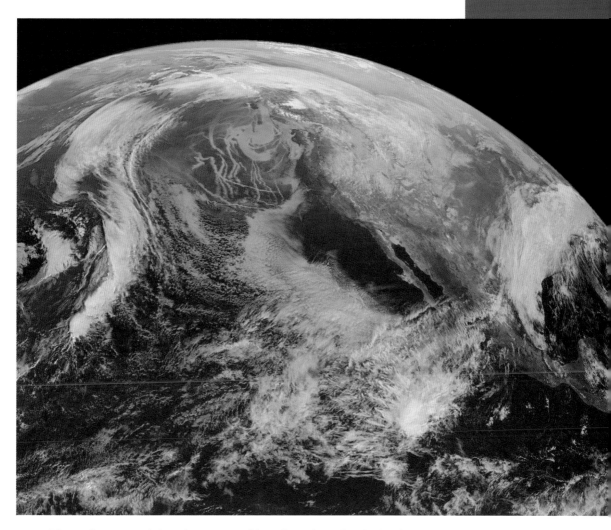

View of the Pacific Ocean with frontal system in mid-latitudes and cumulus cloud in the tropical eastern Pacific.

THIS CHAPTER discusses what is termed climate dynamics – the dynamic and thermodynamic processes that give rise to the large-scale atmospheric circulation. It describes the overall general circulation of the atmosphere, discusses what causes it and how it is maintained, and briefly considers its latitudinal and longitudinal characteristics. In Section 3.1 we considered the basic characteristics of pressure and winds. We begin here by discussing the factors responsible for the general circulation.

5.1 Factors

There are two principal factors that determine the general circulation of the Earth's atmosphere. They are the latitudinal differences of energy received by the Earth–atmosphere system from the sun, with an excess in the Tropics and a deficit in high latitudes, and the corresponding global distribution of angular momentum of the atmosphere. We shall examine these two factors in turn.

The Earth's energy balance involves incoming solar radiation (S) and net outgoing thermal infrared radiation (I), which together make up the net radiation (Rn) (see Section 2.1):

$$Rn = S(1 - \alpha_p) + (I\downarrow - I\uparrow),$$

where α_p = planetary albedo (~ 0.3); $I\downarrow$ = downward infrared radiation; and $I\uparrow$ = upward infrared radiation.

The net radiation is positive equatorward of about 35° latitude and negative poleward of that latitude (see Figure 2.4), thus requiring a poleward transport of energy to maintain radiative equilibrium at each latitude. The poleward transport is a maximum of about 6 PW (PW = petawatt = 10^{15} W) at around latitudes 40°. In the Northern Hemisphere, about two-thirds of this transport takes place in the atmosphere – as sensible heat and latent heat – and about one-third in the ocean, via warm currents. The ocean heat flux is about 2 PW at 20° N, indicating the significance of the Gulf Stream and Kuroshio currents (see Section 8.1), and in the Southern Hemisphere it remains about 1 PW from 20° to 50° S. The atmospheric transport peaks at just over 4 PW at latitudes 40° N and S (Figure 5.1). Sensible heat is transported in the middle and upper troposphere, where the winds are stronger, and latent heat in the lower troposphere where the water vapor resides. This might imply that the mechanisms involved are different. We shall return to this later.

Angular momentum is the product of a body's moment of inertia (a measure of an object's resistance to changes in its rotation rate) and its angular velocity. In other words, it consists of the spin of a body. Angular momentum is conserved if there is no external torque (or twisting motion about an axis). Hence, a spinning ice skater increases his/her rotational speed as the skater's arms are contracted towards the body, closer to the axis of rotation.

For unit mass, the absolute angular momentum of the atmosphere about the Earth's axis of rotation is:

$$M = M_\Omega + M_R,$$

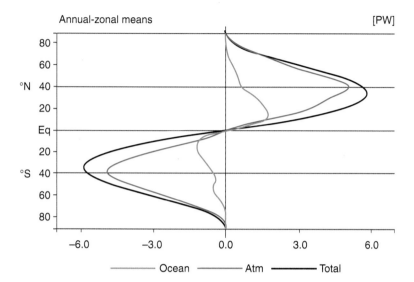

Figure 5.1
The poleward transport of energy with latitude in the atmosphere (red dashed) and ocean (blue dashed) and total (solid) (source: Fasullo and Trenberth 2008).

where $M_\Omega = \Omega r^2 + \cos^2 \phi$ is the angular momentum of the Earth, while

$$M_R = ur \cos^2 \phi$$

is the angular momentum of the atmosphere where Ω (omega) = Earth's angular velocity (7.29×10^{-5} rad s^{-1}), r = Earth's radius (3.67×10^6 m), ϕ (phi) = latitude angle, and u = zonal (westerly) wind speed. (There are 2π radians in a full circle, so one radian (rad) = 57.3 degrees.)

In the Tropics the easterly trade winds gain westerly relative angular momentum from the Earth's rotation and, in the mid-latitude westerlies, the winds lose this westerly angular momentum to the earth by frictional drag of the air over mountain ranges (form drag). Hence, there has to be a momentum transfer from low to middle and high latitudes, otherwise the westerlies would cease within about ten days.

For a stationary body of air at latitude 30°, that is transferred 10° poleward, the air gains 34 m s^{-1} of westerly motion due to conservation of angular momentum. This concept supports the observed increase in westerly wind speeds in the upper troposphere in mid-latitudes (i.e., the jet stream). However, this process does not operate in higher latitudes, where instead the air conserves its absolute vorticity (the sum of the Earth's vorticity and the relative vorticity of the air) that increases from the Pole toward the jet stream (see Section 5.3). The maximum zonal-mean annually averaged total transport of angular momentum is located about 30° N and 40° S. The mechanisms involved in the transport of energy and angular momentum are discussed below.

5.2 Meridional cells and zonal winds

It was first recognized by George Hadley in 1735 that air in the tropical trade winds flows toward the Equator, and as it does so it turns to the right (left) in the Northern (Southern) Hemisphere due to the Earth's rotation, giving rise to north-easterly (southeasterly) trades, respectively. The air is heated at the Equator and rises, moving northward (southward) aloft in the respective hemispheres. This air cools as it moves poleward and sinks back to the surface at about 30° latitude, forming the subtropical high-pressure belts. These are permanent features that shift latitudinally and longitudinally with season. The meridional (north–south), thermally direct cells – where warm air rises and cool air sinks – are called *Hadley cells* (Figure 5.2) in Hadley's honor. The thermally indirect north–south Ferrel cells (named after William Ferrel) in mid-latitudes are very weak and of little conse-quence for transporting energy or momentum (see Box 5A.1).

In the mid nineteenth century, air motion in middle and high latitudes was identified to be mainly horizontal and to move in circular paths around low- and high-pressure systems, but the significance of these motions for transporting energy and momentum was not recognized until the mid twentieth century. In the 1920s, A. Defant suggested that horizontal eddies (low- and high-pressure systems) could transport energy poleward, but it was not established observatio-nally until the 1950s, when V. P. Starr and R. M. White analyzed newly available hemispheric wind and temperature data for the troposphere. They separated

Figure 5.2
Schematic diagram of the global wind belts and the vertical structure of the meridional cells (source: NASA).

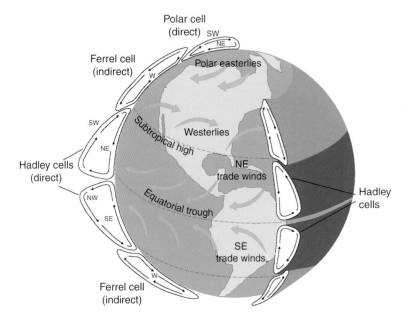

Box 5A.1 Conceptual models of the general circulation

The earliest attempt to suggest a mechanism for the general circulation of the atmosphere was introduced by Edmond Halley in 1685. He indicated a thermally direct meridional cell with maximum heating in low latitudes. Air rises and flows poleward and sinks as it cools, to return equatorward (Figure 5.3a). In 1735, George Hadley improved on this by noting the effects of the Earth's rotation in deflecting the flows to the right in the Northern Hemisphere (Figure 5.3b). However, this scheme overlooked the westerly wind belts. A more complete picture was provided in 1856 by William Ferrel, who proposed a three-cell model similar to that of C.-G. Rossby in 1941, with two direct cells bounding an indirect one in middle latitudes (Figure 5.3c). Ferrel advocated the role of angular momentum conservation to account for the westerlies. The problem of strong upper westerlies in middle latitudes was only recognized in the 1940s. It is important to note that the indirect mid-latitude cell implied upper easterlies rather than westerlies. The solution came in the 1950s with the recognition of the role of horizontal eddies (short waves) in transferring energy and angular momentum poleward.

Nowadays, the view is of a Hadley cell in low latitudes and horizontal eddies in mid-latitudes. In some locations and seasons there may be a small direct cell in high latitudes.

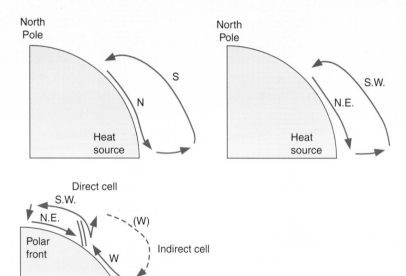

Figure 5.3
Early conceptual models of the general circulation: (a) E. Halley, 1685; (b). G. Hadley, 1735; (c) C.-G. Rossby, 1941 (source: Barry 1967).

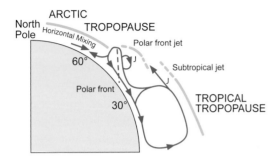

Figure 5.4 The general circulation according to Palme (1951). J denotes a jet stream core, with the wind blowing into the page (source: redrawn according to Palmén, 1951, from Barry 1967).

the poleward transports into meridional cells, stationary planetary waves and transient eddies (cyclones and anticyclones at the surface and short-wavelength traveling waves aloft). They showed that in mid-latitudes horizontal eddies accomplished most of the poleward transport of energy and angular momentum. This view is summarized in Erik Palmén's famous diagram (Figure 5.4). The structure of both planetary and short traveling waves favors poleward transport of energy and momentum because the waves are not symmetrical. Instead, they tilt forward towards the east, so that the poleward limb is longer than the equatorward limb, allowing net energy and westerly angular momentum to be transported poleward. Around the hemisphere there are typically 2–4 almost stationary long waves and 6–12 short waves, superimposed on them, traveling eastward. The vertical Hadley cells in low latitudes and horizontal wave motions in middle and higher latitudes therefore accomplish the poleward transport of energy and angular momentum in the atmosphere. The Hadley cells are more or less symmetrical about the Equator at the equinoxes, but in the summer and winter seasons a single cell is present with a rising arm over the main heat source and sinking in the opposite hemisphere. In the northern summer the southern (winter hemisphere) cell is about five times stronger than the northern one, with cross-equatorial flow. However, most of this asymmetry is in the Asian sector 40–150° E, due to the monsoon, and over the remainder of the globe Hadley cells are present in both hemispheres. The vertical pattern of the poleward transport of sensible and latent heat implies that cyclones and anticyclones transfer the latent heat at low levels and traveling waves transport the sensible heat at upper levels, poleward of the Hadley cells.

5.3 Zonal wind belts

The pattern of global winds at the surface is arranged in three zonal (west–east) belts in each hemisphere (see Figure 3.1). There are the easterly trade winds in the Tropics separated from the mid-latitude westerlies by the subtropical high-pressure belt. The westerlies blow around the equatorward side of the subpolar low-pressure systems, which have on their polar margins a small belt of polar easterly winds. These easterlies are best developed around Antarctica. In the Arctic there is no permanent high pressure, although there is an anticyclone over the Canadian Arctic in spring.

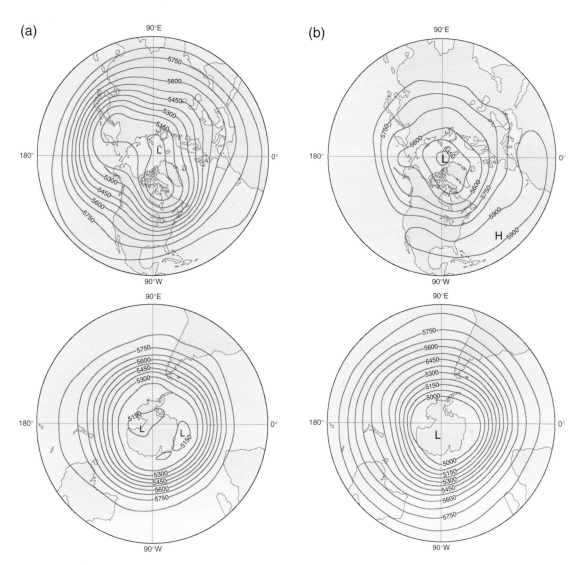

Figure 5.5 The height contours of the 500 mb pressure surfaces (gpm) in each hemisphere for (a) January north and south hemisphere and (b) July north and south hemisphere (source: Barry and Chorley 2010).

In the middle and upper troposphere the wind pattern is much simpler. Over both Poles there is a deep polar vortex with extensive westerlies blowing around it. The subtropical high-pressure centers extend upward into the troposphere and in the Tropics there is a band of weak easterlies extending from the surface into the troposphere. Figure 5.5 shows the 500 mb contours for January and July in each hemisphere.

In the Southern Hemisphere the westerlies are mainly zonal, but in the Northern Hemisphere, in winter, there are three major planetary waves (or long waves) around the hemisphere. This is termed a wave number-three pattern, where the waves have an average spacing of 120° longitude. The wave troughs are located over eastern Asia, eastern North America, and eastern Europe. In summer, there is a single well-developed wave over eastern North America. The waves in winter are attributed to the combined effect of topography (the Tibetan Plateau and the Rocky Mountains) and strong land–sea temperature differences on the atmosphere in the Northern Hemisphere. These tropospheric waves represent the upper level continuation of the surface low-pressure centers over Iceland and the Aleutian Islands that tilt westward with height into the cold air.

Before we can consider the effects of mountain barriers on airflow, we must introduce the concept of *vorticity*. This is the rotation of an air parcel. It can arise from motion in a curved path around a low- or high-pressure center, giving rise, respectively, to cyclonic or anticyclonic relative vorticity. Relative vorticity can also result from differences in wind speed (horizontal shear) across an air current. The two effects may strengthen one another or cancel out. Cyclonic (anticyclonic) vorticity results when, in the Northern Hemisphere, the air is moving faster on the left (right) side of the current, viewed downwind. The patterns are opposite in the Southern Hemisphere. Cyclonic relative vertical vorticity – i.e., that about a local vertical axis – is defined to be positive (anticyclonic relative vorticity is negative). In the Northern Hemisphere, positive relative vorticity is in the same sense as the Earth's rotation.

Westerly airflow crossing a mountain range undergoes anticyclonic (clockwise in the Northern Hemisphere) curvature over the ridge, and cyclonic (anticlockwise) curvature downstream, as a result of the conservation of potential vorticity:

$$\frac{(\zeta + f)}{D} = K,$$

where ζ (zeta) = relative vertical vorticity (cyclonic positive), f = the Coriolis parameter ($2\Omega \sin \phi$ where Ω (omega) = Earth's angular velocity and ϕ (phi) = latitude angle; f = zero at the Equator and 2Ω at the Pole) representing the effect of the Earth's rotation, D = the depth of the air column, and K = a constant. On the windward slope of the range, the vertical air column shrinks (i.e., D decreases), leading to a decrease in ζ (more anticyclonic) in order for the potential vorticity to remain constant. On the lee slope the reverse occurs. Hence, a ridge/trough pattern forms on a continental scale. The trough downstream of the Rocky Mountains is over eastern North America, that downstream of the Tibetan Plateau is over eastern Asia.

These long-wavelength, or planetary, waves in the middle troposphere were first described by C.-G. Rossby in 1939 and hence are often referred to as Rossby waves. Typically there are between two and four such long waves in the Northern Hemisphere westerlies. They tend to move slowly westward, whereas shorter waves (wave numbers 6–12 around the hemisphere) move eastward and are associated with cyclones at the surface. Rossby waves arise as a result of the latitudinal variation of the Coriolis parameter, known as the beta effect. This operates through the tendency for *absolute vorticity* ($f + \zeta$) to be conserved in large-scale flow. Thus, for air moving poleward, f will increase, necessitating that the relative cyclonic vorticity, ζ, decreases. This implies that the air curves anti-cyclonically, which means it returns equatorward. Hence, a planetary wave ridge is formed. As the air moves equatorward, the opposite occurs with f decreasing, leading ζ to increase and forming a trough in the airflow.

The land–sea effect involves the development of a low-level baroclinic (frontal) zone due to the temperature contrast (between cold land and warm ocean) and this is associated with an upper trough in the cold air to the west. In winter the orographic and thermal effects combine. In summer there is only a single plane-tary wave over eastern North America, mostly due to baroclinic effects.

The westerlies of the Northern Hemisphere are much weaker in summer than in winter, but in the Southern Hemisphere there is almost no seasonal change due to the presence of Antarctica. The westerly wind belts have within them the sub-tropical and polar front jet streams, as described in Section 3.1b. Consequently, in the westerly wind belt wind speeds normally increase with altitude up to the top of the troposphere.

The hemispheric westerly circulation alternates at irregular intervals of about 3–6 weeks between a state where the hemispheric flow is strongly zonal and one where it breaks down into a cellular, or *blocking pattern*. This blocking by a persistent high-pressure system disrupts the normal west–east movement of mid-latitude cyclones and causes anomalies of weather conditions to occur over a wide region. Blocking is most common in the winter half-year, and is more developed in the Northern Hemisphere because of the hemispheric long-wave structure that is attributable to the land–sea temperature contrasts and the top-ography of the northern continents. The most common locations of blocking highs are in the northeastern Atlantic, over Alaska and the northeast Pacific, over the Ural Mountains, and over northeastern Canada. In the Southern Hemisphere, where there is less land–sea temperature contrast, the primary area of blocking is over New Zealand. Blocking may set up a ridge extending poleward from one of the subtropical anticyclones, or it may form a separate high-pressure center and cause a split in the jet stream and storm paths, with some systems going north and others to the south, around it. A cut-off low may form in lower latitudes in association with the blocking high. Figure 3.18 shows where many of these tend to occur. Blocking patterns may persist for a month or more and so are important components of regional anomalies of weather and climate.

5.4 Zonal circulations

In the Tropics there is a series of three major east–west (or zonal) vertical circulation cells. These have an upward arm over the heat sources in each of the three equatorial continents – South America, Africa, and the "maritime continent" of Malaysia, the islands of Indonesia and the Philippines (Figure 5.6). Each of these locations is the locus of a heat low, providing strong convection. In the upper troposphere, the air diverges, flowing east and west to descend over the subtropical anticyclones in the eastern parts of the oceans. These cells are known as *Walker circulations*, named after Sir Gilbert Walker who first identified the Southern Oscillation (see Section 6.2). They play a major role in tropical weather regimes. Data on the SST gradient across the equatorial Pacific Ocean indicate that it strengthened during the twentieth century, resulting in an enhancement of both

Figure 5.6
The Walker circulations in a west–east height cross-section through the global equatorial zone (source: after Wyrtki 1985).

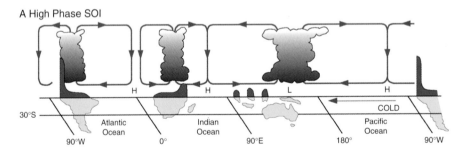

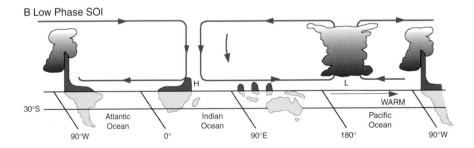

the Walker circulation and the Hadley cell in the Pacific. The upper and lower parts of Figure 5.6 show two different modes of the Walker circulation (see Section 6.2) identified as strong and weak phases of the Southern Oscillation Index (SOI).

SUMMARY

The two main drivers of the global circulation are the latitudinal gradient of net radiation that causes poleward energy transport in the atmosphere and ocean and the corresponding gradient and transport of angular momentum in the atmosphere. The maximum transports of both properties occur about latitude 40° in each hemisphere. The total energy transport maximum is about 6 PW, of which ocean currents account for about one-third at 20° N.

The easterly winds in the Tropics gain westerly relative angular momentum from the Earth rotating beneath them, and the westerly winds in mid-latitudes impart westerly angular momentum to the Earth by form drag over mountains and surface friction. Hence, westerly angular momentum has to be transferred poleward to maintain the westerlies. In low latitudes the transfers of energy and angular momentum are primarily by the vertical plane south–north Hadley cells, with air rising in the Tropics and the subtropical anticyclones located beneath their descending arms. In middle latitudes the transfers are by horizontal eddies – traveling and stationary waves in the middle troposphere. There are typically 2–4 stationary Rossby waves around the hemisphere as a result of mountain ranges and land–sea thermal contrasts. The effect of mountain ranges involves the conservation of potential vorticity as air flows over the barrier. There are about 6–12 short waves around the hemisphere.

The zonal wind belts are easterly in the Tropics (the trade winds) and westerly in middle latitudes, between the subtropical highs and the subpolar lows. Wind speeds increase with height in the westerlies, with a strong jet stream in the upper troposphere. Typically, there is a subtropical jet stream and a polar front jet stream. In summer, over the Indian Ocean–African sector there is a high-level tropical easterly jet stream as a result of upper-level heating over the Tibetan Plateau.

The mid-latitude westerlies sometimes break down into a cellular pattern with blocking high-pressure cells. These patterns are more common in the Northern Hemisphere; the Southern Hemisphere circulation is more zonal.

Along the Equator there are three troposphere-deep standing zonal circulations – Walker circulations – with rising arms over the three heat sources in Amazonia, the Congo, and the maritime continent, and sinking over the eastern oceans. At irregular intervals the Pacific cell is displaced eastward.

QUESTIONS

1   Describe the circulation mechanisms that transport energy and angular momentum poleward.
2   Explain the changing role with latitude of the oceans and atmosphere in transporting energy poleward.
3   Describe the role of potential vorticity conservation in setting up planetary waves.
4   Compare Hadley cells and Walker cells.
5   What mechanisms account for the quasi-stationary trough in winter over eastern North America?
6   Compare the characteristics of long planetary waves and short waves.
7   Describe the mid-latitude circulation when there is blocking.
8   Why is blocking uncommon in the Southern Hemisphere?
9   Locate a series of daily or monthly weather maps (e.g., the weather log published each month in *Weather*) and categorize the circulation as zonal or blocked. Examine the temperature and precipitation anomalies with each pattern.

Circulation modes

6

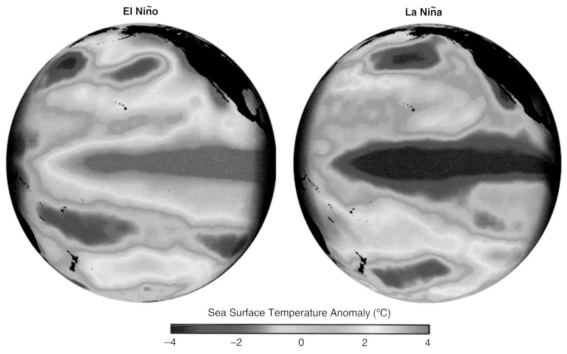

El Niño   La Niña

Sea Surface Temperature Anomaly (°C)

−4   −2   0   2   4

Global maps centered on the Pacific Ocean showing contrasting patterns of sea surface temperature anomalies (°C) during El Niño and La Niña.

6.1 Introduction

It has been known since the work of Sir Gilbert Walker in the 1920s that there are major large-scale oscillations within the general atmospheric circulation. Walker described the famous *Southern Oscillation* in sea-level pressure across the equatorial Pacific Ocean, the North Atlantic Oscillation between the Azores high-pressure cell and the Icelandic low, and the North Pacific Oscillation between the North Pacific high and the Aleutian low. Each of these involves a seesaw in the pressure gradients between the respective centers. It has subsequently been shown that there are many others, including the Northern Annular Mode (NAM) and the Southern Annular Mode (SAM) between northern and southern middle and high latitudes, respectively. The identification of these oscillations became possible when there were extensive series of daily pressure and geopotential height data. Now indices of the principal oscillations are routinely updated by NOAA. In the North Pacific and North Atlantic there are long-term oscillations in sea surface temperatures (SSTs).

Different modes of these oscillations can persist for months to years and they exert a major influence on regional and global climate, modifying storm tracks and the distribution of climatic anomalies.

6.2 The Southern Oscillation and El Niño

The Southern Oscillation is a fluctuation of about ± 2 mb in surface air pressure between the equatorial eastern and the western Pacific Ocean. Its strength is measured by the difference in MSL pressure between the subtropical high over Tahiti (18° S, 150° W) and the equatorial low over Darwin, Australia (12.5° S, 131° E), referred to as the Southern Oscillation Index (SOI). The inverse correlation coefficient between these two remote pressure centers is about –0.8, meaning that when pressure is high in one center it is low in the other (Figure 6.1). This phenomenon of distant correlations in atmospheric conditions is known as a *teleconnection* (see Box 6A.1). There are many such teleconnections in the atmosphere.

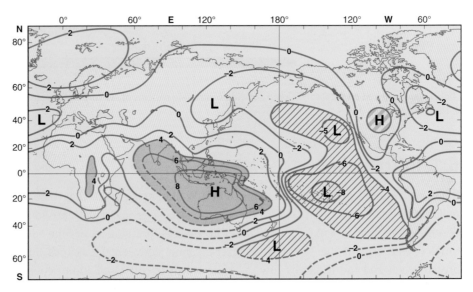

Figure 6.1 Simultaneous correlations (×10) of mean sea level pressure with that at Darwin, Australia (source: Trenberth and Shea 1987).

Box 6A.1 **Teleconnections**

Between 1909 and 1930 Sir Gilbert Walker described correlations between atmospheric pressure, temperature, and rainfall anomalies at great distances (3000–6000 km). However, some of these correlations were later found not to be robust over time and interest waned. In 1935 A. Angstrøm proposed the term teleconnection for spatial patterns of climatic fluctuations. The term was adopted by J. Bjerknes in 1969 to refer to patterns of circulation in the equatorial Pacific. Renewed attention was given to them in the 1980s and 1990s, when a large number of patterns was identified. In winter months in the Northern Hemisphere correlations range from 0.72 to 0.86. Figure 6.2 illustrates the five most prominent in the middle troposphere. A variety of statistical methods are now used to identify the patterns.

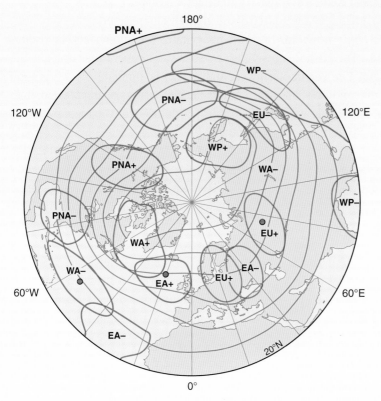

Figure 6.2 Simultaneous correlation patterns in winter 500 mb heights for 1962/1963–1978/1979. Centers of the five strongest patterns. EU Eurasian, WP Western Pacific, PNA Pacific North American, WA West Atlantic, EA East Atlantic (source: Wallace and Gutzler 1981).

The Southern Oscillation in the atmosphere has a time scale of the order of 2–3 years. It is closely coupled to variations in tropical Pacific SSTs. Typically, when the SOI is large, with stronger high pressure in the eastern tropical Pacific, the eastern ocean is cool with strong oceanic upwelling, and the western Pacific is warm. The upwelling in the eastern Pacific is caused by persistent easterly trade winds pushing the surface water away from the coast of South America.

At about 3–7-year intervals, the trade winds weaken and the eastern Pacific warms, often around Christmas, and the pattern of vertical zonal circulation cells over the Pacific Ocean shifts eastward as pressure rises over Indonesia. The warm southward-flowing ocean current off Peru is called *El Niño*, Spanish for the Christ child. The opposite cool mode is known as La Niña, the girl, associated with the cool north-flowing Humboldt (or Peru) Current. These two phases of the coupled El Niño–Southern Oscillation (ENSO) pattern are shown in Figure 5.6, and the ocean conditions of each in Figure 6.3 and Chapter 6 plate.

El Niño and La Niña are the source of the greatest climatic anomalies on Earth. In Peru, with El Niño conditions there are heavy rains that often cause landslides in the mountains. The offshore upwelling ceases, water temperatures rise, and the fisheries collapse. The Pacific northwest and Alaska is anomalously warm, and there are drought conditions in eastern Australia, southern Asia, and Mexico. In El Niño years there are fewer west Pacific tropical cyclones over Guam, and there is also cyclone activity in the tropical Atlantic.

During La Niña there are strong easterly trade winds across the tropical Pacific Ocean, and in Indonesia there is low pressure and strong vertical convection giving rise to wet weather. The low-level easterlies and vertical motion generate an upper-level westerly return flow that makes up one cell of the so-called west–east (zonal) Walker circulation. There are two other corresponding cells over Africa and the Indian Ocean, and South America and the Atlantic Ocean (see Figure 5.6). The persistent strong easterlies over the Pacific Ocean raise sea level in the western Pacific Ocean by about 60 cm as a result of wind stress on the ocean surface. La Niña patterns give rise to wet conditions in the Pacific Northwest of North America and dry conditions in the southwest and southeastern United States and in the west-central Pacific. Northeastern South America tends to be wet.

Both ENSO modes appear to be closely linked to the annual cycle, tending to develop in boreal spring and lasting at least a year. There appears to be a close coupling between the atmosphere and ocean. In March–April there is a peak in the occurrence of low-level westerly wind bursts (WWBs) lasting 7–10 days in the western equatorial Pacific. Many of these WWBs seem to result from the simultaneous presence of tropical cyclones in each hemisphere, giving westerlies along the Equator. These WWBs set up very long wavelength ocean Kelvin waves that cross the Pacific to the east in 3–4 months and lead to warming off South America.

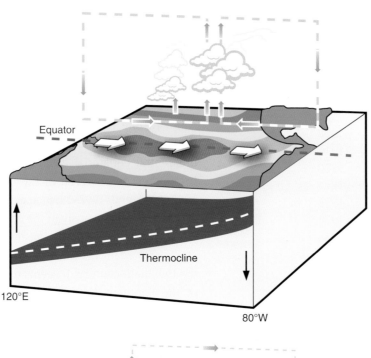

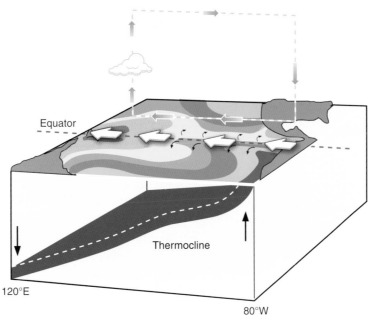

Figure 6.3
The oceanic conditions during (a) El Niño and (b) La Niña in the tropical Pacific (source: Wikipedia, http://en.wikipedia.org/wiki/El_Niño-Southern_Oscillation).

A further forcing mechanism for ocean Kelvin waves is the Madden–Julian Oscillation described in Section 6.11.

Table 6.1 shows the occurrence of warm and cold events in the tropical Pacific since 1900. For 1950–1997, K. Trenberth identified 15 events with 31 percent of

Table 6.1 Years of warm and cold events in the tropical Pacific, 1901–2011

Warm	Cold
1902	1903
1904	1906
1911	1908
1913	1916
1918	1920
1923	1924
1925	1928
1930	1931
1932	1938
1939	1942
1951	1949
1953	1954
1957	1964
1963	1970
1965	1973
1972	1975
1976	
1982	
1986	1988
1991	
1992	1995
1997	1998
2002	2003
2006	2007
2009	2010

The year listed refers to the start of the event. After van Loon (1984), Diaz and Kiladis (1992); updated from NOAA.

months as El Niño, and ten events with 23 percent of months as La Niña. It is important to note that neither mode was present for 45 percent of the time. During these intervals the SST anomalies across the equatorial Pacific are weak.

There appear to be two different "flavors" of El Niño. In one pattern the warming begins along the west coast of South America and spreads into the central equatorial Pacific in austral summer. In another, as observed in the intense events of 1982–1983 and 1997–1998, maximum anomalies occurred simultaneously in the eastern and central equatorial Pacific during austral summer. The 1997–1998 event has been termed the El Niño of the twentieth century. It seems that the properties of the Southern Oscillation changed during the 1980s and 1990s according to Federov and Philander. La Niña episodes were very weak or practically absent from 1976–1994 (see Table 6.1), whereas El Niño attained unprecedented amplitudes in 1982 and 1997 and was unusually prolonged in 1992. The changes are attributed either to random fluctuations or to global warming effects. A decision on which is the case has not yet been reached.

From the western end of the oscillation in the tropical Pacific, wave energy in the atmosphere propagates northeastward into the westerly wind belt. Commonly, a *Pacific–North America (PNA) pattern* (see Figure 6.2) of a ridge and troughs is set up, with the troughs over the eastern North Pacific and eastern North America, and the ridge over the western United States (positive PNA). The positive mode is associated with above-average precipitation in Alaska and the Pacific Northwest and drier conditions over the Midwestern United States. This pattern is typically reversed (negative PNA) with La Niña. In this case cyclones off East Asia track northeastward into the Bering Sea, with a second cyclone area off western Canada. The PNA pattern has a barotropic structure and is a major source of climate variability over North America during autumn, winter, and spring; it is absent in June and July.

6.3 The North Atlantic Oscillation

The *North Atlantic Oscillation* (NAO) is an oscillation in sea-level pressure between the Azores subtropical anticyclone and the Icelandic low-pressure center (Figure 6.4). When the pressure gradient is large the NAO index is positive and there are strong westerlies, and when the gradient is reversed the index is negative and the westerlies are disrupted. The NAO is the dominant mode of winter climate variability in the North Atlantic sector. During extreme negative NAO events there are typically well-below-normal temperatures in northwestern Europe and above-normal precipitation in the western Mediterranean. Also, there are above-normal temperatures in western Greenland and below-normal temperatures in the eastern United States. The NAO determines the strength and direction of the mid-latitude westerlies and storm tracks across the North Atlantic.

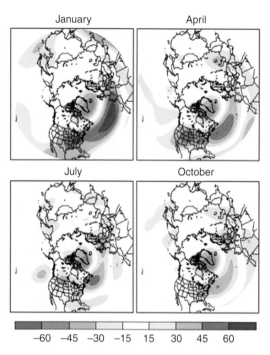

Figure 6.4 The North Atlantic Oscillation in its positive mode for mid-season months (source: NOAA, www.cpc.ncep.noaa.gov/data/teledoc/nao_map.shtml).

In positive NAO winters there is more precipitation over northern Europe and the Mediterranean is drier; the opposite occurs in negative NAO winters. The NAO index varies in intensity and sign, but does not exhibit any periodicities. The index was largely positive from 1900 to 1950, negative in the 1960s, and has been mostly positive since 1970, until about 2006.

6.4 The Northern Annular Mode

The *Northern Annular Mode* (NAM), also known as the Arctic Oscillation (AO), involves low pressure over the Arctic Ocean and high pressure in mid-latitudes (positive mode) or vice versa (negative mode) (Figure 6.5). It is represented throughout the troposphere. Thompson and Wallace first described it and showed that high index (strong NAM) days feature westerly surface winds along 55° N and transpolar flow from Russia toward Canada, whereas low-index conditions are marked by cold anticyclones centered over central Canada and Russia and an anticyclonic surface circulation throughout the Arctic basin. High-index days are, on average, 5 °C warmer over much of the Midwestern United States, central Canada, Europe, and the Barents–Kara seas. Cold events occur with much greater

Negative phase

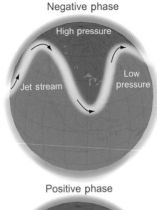

Positive phase

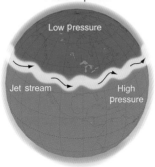

Figure 6.5 The negative (top) and positive (bottom) phases of the Arctic Oscillation or Northern Annular Mode (source: Wikipedia, http://en.wikipedia.org/wiki/Arctic_oscillation).

frequency over North America, Europe, Siberia, and East Asia under low NAM index conditions. The NAM was mostly slightly negative from 1950–1990 and then mostly positive until 2008.

6.5 The Southern Annular Mode

The *Southern Annular Mode* (SAM), or Antarctic Oscillation (AAO), is the Southern Hemisphere counterpart of the NAM, with opposing pressure and geopotential height anomalies over Antarctica and southern middle latitudes. The index was mostly negative from 1980 to 1993, positive to 2000, then negative to 2008.

There is an increasing trend in the SAM in the austral summer since the mid 1960s, and in austral autumn since 1958, that has been linked to the intensification of the ozone hole in the stratosphere.

6.6 The semiannual oscillation

The semiannual oscillation is a feature of the mid-tropospheric circulation in the Southern Hemisphere. The mechanism involves the different annual cycles of air temperature between the Antarctic and the surrounding mid-latitude southern oceans. The temperature difference is strong at the equinoxes, particularly the autumn one in March. At latitude 50° S cooling in autumn is rapid compared with the warming in spring (September), the opposite being the case at latitude 65° S. This behavior is related to the heat budget of the upper ocean. The semiannual maximum temperature gradients in the middle troposphere over the Southern Ocean, through increased cyclonic activity, shift the circumpolar trough poleward during the transition seasons. In turn, this produces a semiannual oscillation in the pressures ($\sim$ 5–6 mb) and winds over the area affected by the trough.

The hemispheric mean pressure in mid-latitudes falls with the equatorward shift of the circumpolar trough from autumn to winter, while the pressure rises over Australia, South America, and southern Africa.

6.7 The North Pacific Oscillation

The *North Pacific Oscillation* (NPO) is the North Pacific counterpart of the NAO. It involves the Aleutian low and the subtropical North Pacific anticyclone. When the pressure gradient between them is pronounced, the mid-latitude westerlies are strong and more zonal across the North Pacific. It has major effects on air temperature in winter in Alaska, the Pacific Northwest, Canada, and the United States, and exceeds the effects of the PNA or ENSO. It also has greater influence on winter precipitation in the Pacific Northwest, western Mexico, and the south-central Great Plains. The NPO has fluctuated in sign over 10–15-year intervals; it was generally positive from the mid 1980s–1990s, and then declined. It is also modulated by the Pacific Decadal Oscillation (see Section 6.8).

6.8 The Pacific Decadal Oscillation

The *Pacific Decadal Oscillation* (PDO) features warm or cool surface waters in the North Pacific Ocean, north of 20° N. During a "warm" (positive) phase, the west Pacific cools up to 0.5 °C and part of the eastern ocean warms; during a "cool" (negative) phase, the opposite pattern occurs. The oscillation is quasi-periodic, each phase lasting for 20–30 years. "Cool" PDO regimes prevailed over 1890–1924, 1947–1976 (except around 1958–1960), and since about 1998, while "warm" PDO regimes dominated 1925–1946 and 1977 through the late 1990s.

The PDO is not a single mode of oceanographic variability, but arises through the combined influence of ENSO, atmospheric variability in the extratropics, and the North Pacific Ocean gyre.

Extremes in the PDO pattern are marked by widespread variations in the Pacific Basin and the North American climate. The PDO has considerable influence on climate-sensitive natural resources in the Pacific and over North America, including the water supplies and snow pack in some selected regions in North America, and major marine ecosystems from coastal California north to the Gulf of Alaska and the Bering Sea. In fact, it was fisheries scientist Steven Hare that coined the term Pacific Decadal Oscillation (PDO) in 1996 while researching connections between Alaska salmon production cycles and Pacific climate. Causes for the PDO are not currently known. Likewise, the potential predictability for this climate oscillation is unknown. Understanding the mechanisms giving rise to PDO will determine whether skillful decades-long PDO climate predictions are possible. Even in the absence of a theoretical understanding, PDO climate information improves season-to-season and year-to-year climate forecasts for North America because of its strong tendency for multi-season and multiyear persistence.

6.9 The Atlantic Multidecadal Oscillation

The *Atlantic Multidecadal Oscillation* (AMO) is a mode of variability of SSTs in the North Atlantic Ocean, 30–65° N, analogous to the PDO. It was first identified in the 1990s. Since about 1860 it has oscillated with a period of about 70 years; there was a warm phase 1930–1960, and it is currently positive again. Its amplitude is about 0.4 °C. There is debate as to how much is externally forced and how much is due to internal variability. The signal of global warming has to be removed from the SST time series before it can be analyzed. Models suggest that the AMO is caused by small changes in the North Atlantic thermohaline circulation. The signal appears to be linked to precipitation anomalies in North America, the summer climate of North America and Europe, and multidecadal variations in droughts in the Sahel. Major droughts in the Midwest and southwest United States were more severe during the AMO warm phase in the 1930s and 1950s. It also has played a role in the formation of severe Atlantic hurricanes – suppressing their activity in the 1970–1990 cool phase.

6.10 Southern Hemisphere wave number-three pattern

In austral winter (May–September) 500 mb geopotential heights over Africa exhibit positive correlations with heights over South America and the central South Pacific near New Zealand, and negative correlations with height over the Southern Ocean. A wave number-three pattern is apparent – with three troughs and three ridges (see Section 5.3), showing the correlation of 500 mb geopotential height anomalies in the Subtropics with 500 mb anomalies at 50° S, 95° E during winter. This wave number-three pattern is one commonly associated with blocking patterns in the Southern Hemisphere. During summer (November–March), anomalies over the three continents occur out of phase with anomalies over the subtropical oceans, and in a wave number-three pattern over the Southern Ocean near 55° S.

6.11 The Madden–Julian Oscillation

The *Madden–Julian Oscillation* (MJO) was discovered by R. Madden and P. Julian in 1971 through an analysis of zonal wind anomalies in the equatorial Pacific. It is a large-scale coupling between atmospheric circulation and deep convection in the Tropics (Figure 6.6). It makes up the largest fraction of the intra-seasonal (30–90-day) variability in the tropical atmosphere. It is a traveling pattern, with the convective center propagating eastwards at 4–8 m s^{-1} over the warm parts of the Indian and Pacific oceans. An area of anomalous rainfall first appears over the western Indian Ocean, and remains evident as it propagates over the very warm ocean waters of the western and central tropical Pacific. The anomaly in convection weakens as it moves over the cooler waters of the eastern Pacific Ocean, but

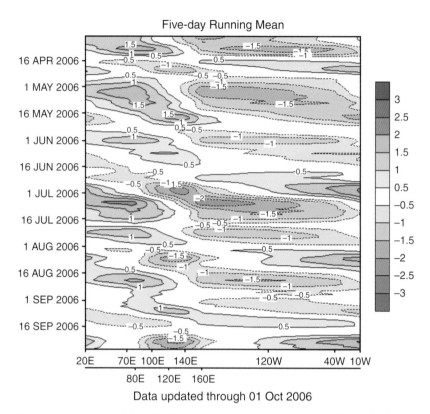

Figure 6.6 Madden–Julian Oscillation in a time–longitude (Hovmöller) plot for 1 April to 1 October 2006. The diagram shows a five-day running mean of outgoing longwave radiation (OLR) anomalies (W m^{-2}), which indicate anomalies of deep cumulonimbus cloud cover. Cold, high cloud tops emit little OLR (source: Wikipedia).

the anomalies in atmospheric circulation continue around the globe. The wet phase is followed by a dry phase where thunderstorm activity is suppressed. Each cycle lasts 30–60 days.

Its cause remains uncertain. A recent idea is that evaporation leads to low-level moisture convergence, which builds up cumulus convection that gradually amplifies into deep convection. It appears that there is meridional moisture convergence to the east of the center of convection.

Teleconnections between MJO and North American climatic anomalies in boreal winter have been shown by Zhou *et al.* When enhanced convection is located over the equatorial Indian Ocean, below-average air temperatures tend to occur in New England and the Great Lakes region. As enhanced tropical convection shifts over the Maritime Continent (Indonesia), above-average temperatures appear in the eastern states from Maine to Florida. The MJO also influences winter precipitation. When enhanced convection is located over the maritime continent, more precipitation falls in the central plains. Enhanced precipitation also occurs over the west coast when the convection center is over the Indian Ocean.

In Figure 6.6 time increases from top to bottom, so contours that are oriented from upper-left to lower-right represent movement of the anomalies from west to east. A full list of atmospheric oscillations and time series of the data are available at www.esrl.noaa.gov/psd/data/climateindices/list

It should be emphasized that there are strong interactions between many of the atmospheric teleconnections and these may operate to strengthen or weaken their effects on regional and hemispheric climatic anomalies.

SUMMARY

There are many large-scale standing pressure oscillations in the atmosphere. The best known is the Southern Oscillation teleconnection pattern in sea-level pressure between the anticyclone in the eastern equatorial Pacific and low pressure over Indonesia. This atmospheric oscillation is closely coupled with the oceanic El Niño–La Niña phenomena, which involve anomalous SSTs in the eastern equatorial Pacific Ocean, respectively, warm and cold. Each phase typically begins in March and may last 1–2 years. These two modes each give rise to pronounced climatic anomalies in many parts of the world and represent the principal source of interannual climatic variability. With El Niño, the Pacific–North America teleconnection pattern is usually in its positive mode, with a strong pressure ridge over western North America and low pressure in the southeastern United States. This gives corresponding positive and negative temperature anomalies.

The North Atlantic and the North Pacific oscillations involve the subtropical high-pressure cells and the subpolar lows in each ocean. High index values of either oscillation represent strong zonal westerlies. The Pacific Decadal Oscillation (PDO) features a warm eastern and cool western North Pacific Ocean in its positive phase and vice versa in its negative phase. Each phase lasts 20–30 years. The Atlantic Multidecadal Oscillation (AMO) is the Atlantic counterpart of the PDO, with an amplitude of about 0.4 °C and a period of about 70 years. Major droughts in the Midwest and southwest United States were more severe during the AMO warm phase in the 1930s and 1950s. The Arctic Oscillation (AO), or Northern Annular Mode (NAM), and the corresponding Antarctic Oscillation (AAO) or Southern Annular Mode (SAM) each involve opposite pressure anomalies over the polar caps and middle latitudes. These patterns extend through the troposphere.

Positive modes give rise to strong zonal circulations. Strengthening of the SAM in summer–autumn over the last five decades appears to be linked to intensification of the Antarctic ozone hole in the stratosphere.

In the Southern Hemisphere there is a wave number-three pattern. In austral winter 500 mb geopotential height anomalies between the Africa, South America, and central South Pacific are positively correlated, and negatively correlated with heights over the Southern Ocean. The is also a semi-annual pressure and temperature oscillation in the Southern Hemisphere. The Madden–Julian Oscillation (MJO) is a traveling pattern of circulation anomalies and convection in the Tropics. It accounts for most of the variability in the Tropics on a 30–90-day time scale. Convection propagates eastward at 4–8 m s^{-1} over warm waters of the Indian and Pacific oceans. A wet phase is followed at 30–60-day intervals by a dry phase. Various phases of the MJO are shown to be linked to temperature and rainfall anomalies in the United States.

QUESTIONS

1 Using the data in Table 6.1 calculate the mean interval between warm El Niño events and plot a frequency distribution.
2 Select one oscillation from the NOAA web site and examine and comment on the time series of its mode.
3 Explain why the various teleconnections are important for global and regional climate.
4 Compare the climatic conditions in North America during a warm ENSO event and a cold event using maps and data from the NOAA web site.
5 Examine the similarities between the NAM and SAM teleconnection patterns.
6 Contrast the nature of the MJO with ENSO.
7 Select years with both positive and negative NAO indices from the NOAA tables and compare and contrast the January climate anomalies in Chicago, New York, and London for those years.
8 Compare the climatic anomalies over western North America during positive and negative phases of the Pacific–North American pattern.

7 Synoptic climatology

Satellite image for 5 September 2004 showing a low pressure system over the Upper Midwest of the United States and Hurricane Francis over Florida.

THE SUBFIELD of synoptic climatology involves the study of local, regional, continental, or hemispheric climatic conditions in the context of atmospheric circulation patterns defined by airflow or pressure fields.

This chapter examines regional classifications, especially the well-known classification of Hubert Lamb for the British Isles, but also the spatial synoptic classification that is applied at individual weather stations. It then considers the European–North Atlantic scheme of *Grosswetterlagen* (large-scale weather patterns). Finally, two hemispheric classifications developed in the former Soviet Union are discussed. The chapter examines the advantages and limitations of both subjective empirical approaches and objective numerical ones. It closes by giving examples of recent applications of synoptic climatology in evaluating the realism of the outputs of general circulation models and in assessing teleconnection patterns.

7.1 Introduction

Synoptic climatology is the study of climate from the perspective of the atmospheric circulation, with emphasis on the connections between regional daily circulation patterns and climatic conditions. This subfield developed during World War II, when efforts were made to predict weather conditions over Japan from patterns of airflow. Major texts on the subject have been written by Barry and Perry in 1973 (*Synoptic Climatology: Methods and Applications*), Barry and Carleton in 2001 (*Synoptic and Dynamic Climatology*), and Yarnal in 1993 (*Synoptic Climatology in Environmental Analysis: A Primer*). The common approach is to determine distinct categories of synoptic weather patterns and then to assess statistically the weather conditions associated with these patterns. A typical classification has about 20–30 types. The patterns may be based on isobaric or geopotential maps or airflow directions. The geographical scales involved range from regional to hemispheric.

Until the 1970s, classifications were developed subjectively; a skilled analyst looked at each day's weather map and assigned it to a type. Then, with the availability of digital files of pressure data and computers, numerical classification methods were developed. These range from correlation-based approaches where pairs of maps are correlated with one another, to empirical orthogonal functions (EOFs) (or principal component analysis) where underlying patterns are extracted statistically, to self-organizing maps where artificial neural networks are used to group isobaric patterns into discrete types. A description of these approaches is provided by Barry and Carleton. A big advantage of objective numerical approaches is that the classification can easily be updated. A comparison and evaluation of various methods that have been applied for Europe is given by Huth *et al*.

There are several weaknesses involved in synoptic classifications. First, the atmosphere is continuous so the determination of any boundary between circulation categories is arbitrary. Second, pressure systems are of varying intensity, affecting their characteristic weather. Third, there are seasonal differences related to the changing intensity of the weather systems and contrasts in land/sea conditions that affect the air masses involved.

7.2 Regional classifications

The best-known regional classification of synoptic circulation types is that developed by Hubert Lamb for the British Isles (see Box 7B.1). For the surface, there are eight directional *airflow types* (N, NE, E, SE, S, SW, W, and NW), each subdivided according to the curvature of the isobars (cyclonic, anticyclonic, or neutral), plus cyclonic and anticyclonic, giving 26 types in all. Originally, each day's weather

Box 7B.1 **Hubert Lamb**

Lamb was born in Bedford, England in 1913. Much of his professional life was spent in the UK Meteorological Office, where he worked on long-range forecasting, world climatology, and climatic change. He was one of the first climatologists to recognize that climate could change on human time scales. In 1950 he published his classification of British weather types and maintained a daily catalog until his death in 1997. In the 1960s he wrote a study on the Medieval Warm Epoch and Little Ice Age based on historical documents and botanical evidence, and in 1970 he published a monograph on volcanoes and climate in which he proposed a Dust Veil Index (DVI). In 1971 he left the Meteorological Office to become the first Director of the Climatic Research Unit (CRU), established in 1972 at the University of East Anglia. He secured funding from insurance companies interested in climatic assessments of storms and floods. He published many notable papers and books on climatology and climate change.

Box 7A.1 **Reanalysis**

Reanalysis, or retrospective analysis, involves the use of a state-of-the-art analysis/forecast system to perform data assimilation using all available quality-controlled past data (surface, ship, upper air, aircraft, and satellite) to generate consistent atmospheric fields for climatological analysis.

A web site listing available reanalysis data sets is at: http://reanalyses.org/atmosphere/overview-current-reanalyses

The most frequently used are the reanalyses produced by the National Centers for Environmental Prediction (NCEP), with the National Center for Atmospheric Research (NCAR), which starts in 1948 and continues through to the present, and the European Centre for Medium Range Weather Forecasts (ECMWF) ReAnalysis ERA-40 over 1957–2002. Several reanalysis products begin in 1979.

map was categorized subjectively, but in the 1990s an objective numerical approach was adopted. The original catalog of daily types extends from 1861 to 1997. The objective catalog, based on methods developed by A. F. Jenkinson and F. P. Collison of the UK Meteorological Office in 1977, makes use of three basic variables that define the surface circulation features over the British Isles: the direction of the mean flow; the strength of the mean flow; and the relative vorticity. It extends from 1880 through the present, and is available at www.cru.uea.ac.uk/cru/data/lwt

Recently, a new catalog has been developed using the same criteria, with reanalyses for 12 Universal Time Coordinated from the new twentieth-century reanalysis (20CR) for 1871–1947, and the National Center for Environmental Prediction (NCEP) from 1948 onwards, by P. Jones *et al.* (see Box 7A.1).

Table 7.1 Characteristics of Lamb's "weather types"

Westerly: Unsettled weather with variable wind directions as low-pressure systems cross the country. Mild and stormy in winter, cool and cloudy in summer.

Northwesterly: Cool, changeable conditions. Strong winds and showers affect windward coasts, but southern Britain may have dry, bright weather.

Northerly: Cold weather at all seasons, often associated with polar lows. Snow and sleet showers in winter, especially in the north and east.

Easterly: Cold in the winter half-year, sometimes very severe weather in the south and east with snow. Warm in summer, dry weather in the west; occasionally thundery.

Southerly: Warm and thundery in summer. In winter may be associated with a low in the Atlantic, giving mild damp weather in the southwest, or with a high over Central Europe, in which case it is cold and dry.

Cyclonic: Rainy, unsettled conditions, often with gales and thunderstorms. This type may refer either to the rapid passage of depressions across the country or to the persistence of a deep depression.

Anticyclonic: Warm and dry in summer, occasional thunderstorms. Cold and frosty in winter with fog, especially in autumn.

A description of the characteristics of the Lamb types in Britain is given by Barry and Chorley. A summary of the climatic characteristics of the major types in the British Isles, called "weather types" by Lamb, is given in Table 7.1.

The three major types (cyclonic, anticyclonic, and westerly, which comprise about 250 days per year [68 percent of days]), do not exhibit any major long-term trends, contrary to earlier conclusions based on the subjective classification of Lamb.

On an annual basis, the most frequent airflow type over the British Isles is westerly; it has a 35 percent frequency in December to January and is almost as frequent in July to September. The minimum occurs in May (15 percent), when northerly and easterly types reach their maxima (about 10 percent each). Pure cyclonic patterns are most frequent (13–17 percent) in July to August and anticyclonic patterns in June and September (20 percent); cyclonic patterns have > 10 percent frequency in all months and anticyclonic patterns > 13 percent.

The effects of changes in circulation type frequency versus within-type changes of climatic characteristics was demonstrated by Perry and Barry for four stations in Great Britain between 1925–1935 and 1957–1967. Temperature changes in January were attributable to changes in the frequency of types, whereas changes in April, July, and October were linked to within-type changes in characteristics.

The equation used to separate the two contributions to temperature changes is given in Barry and Carleton.

For the central United States, Coleman and Rogers developed a classification of ten types using both surface and upper-air data for 1948–2004. They first determined empirical orthogonal functions and then performed a clustering analysis to obtain the types. Three types occur in both summer and winter and four types in winter and the transition seasons. The annual frequencies of two winter synoptic types, associated respectively with strong zonal and meridional flow, are highly correlated with the phase of the Pacific–North American teleconnection pattern (see Section 6.2), while eastern equatorial Pacific sea surface temperatures (SSTs) are linked to a synoptic type featuring low pressure around the Gulf Coast.

For the Arctic of North America, a classification has been developed by Cassano *et al.* using *self-organizing maps* (SOMs). The SOM technique uses a neural network algorithm that employs unsupervised learning to determine generalized patterns in data – in this case isobaric pressure patterns (Figure 7.1). For the chosen region, 35 types are recognized. The temperature conditions in Alaska are determined for each SOM pattern. It is also shown that over 80 percent of the warming that occurred around 1976 was not attributable to circulation changes that took place over the North Pacific at that time. The warming could be part of a global signal, related to higher SSTs off northwest Canada, or due to increased nighttime cloud cover.

A comparison of different circulation type schemes for Europe has been carried out by Philipp *et al.* For 1957–2002, daily mean sea level (MSL) pressure data were classified using predefined types, including manual and threshold-based classifications, and methods that produced types derived from the input data, including ones based on EOFs and clustering techniques, key pattern algorithms, and optimization algorithms to identify clusters, such as SOM. Seventeen different approaches were analyzed, and results show that subjective classifications compared to automated methods show higher persistence, greater inter-annual variation, and long-term trends. Among the automated classifications, optimization methods show a tendency for longer persistence and higher seasonal variation. Overall, there is better performance in winter months and for smaller domains located in western Europe.

A different approach – the spatial synoptic classification (SSC) originally developed by Kalkstein and Corrigan – is adopted by Sheridan based on daily air mass types at individual weather stations in the United States and Canada. Six types are defined: (1) dry polar (DP); (2) dry moderate (DM); (3) dry tropical (DT); (4) moist polar (MP); (5) moist moderate (MM); and (6) moist tropical (MT). Maximum and minimum temperatures, dew point temperature at 1600 h and mean daily cloud cover are analyzed. "Seed days," which contain the typical characteristics of each

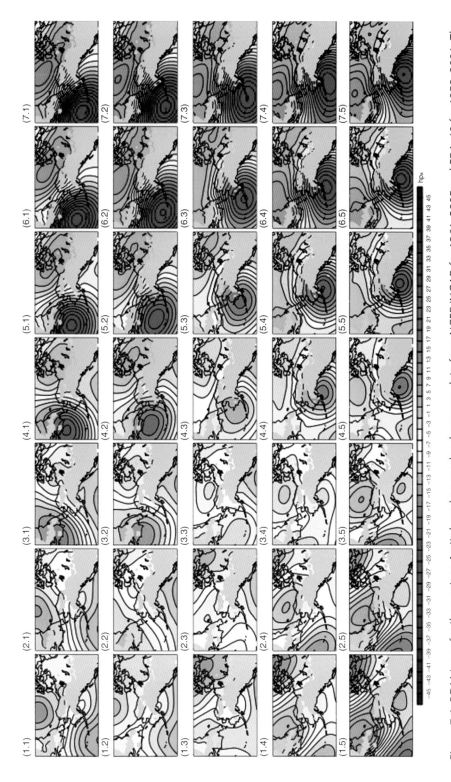

Figure 7.1 SOM types for the western Arctic based on sea-level pressure data from NCEP/NCAR for 1961–2005 and ERA-40 for 1958–2001. The contour interval is 2 mb and the color scale shows a range of anomalies from −45 to +45 mb (source: Cassano et al. 2011).

air mass type, are identified for any season of the year as the basis for classifying all days. The daily calendars for 327 stations in North America, since 1948 for the United States and for 1953–1993 for Canada, are available at http://sheridan.geog.kent.edu/ssc.html

7.3 Continental classifications

For the northeastern North Atlantic and Europe, the *Grosswetterlagen* (large-scale weather pattern) classification, originally developed by Franz Baur in the 1930s–1940s and extended by P. Hess and H. Brezowsky in 1952, is regularly updated in a computerized version and is still widely used. It is based on 29 circulation patterns at the surface and 500 mb levels. In summary, there are four west types, two northwest, six north, two northeast, four east, two southeast, four south, two southwest, two Central European high patterns and one Central European low. There is an extensive literature on the climatic characteristics of the Grosswetter patterns at European stations. Modified Grosswetter classifications have also been developed for the Alps.

From an examination of the duration of circulation patterns since 1881, Kisely and Domonkos identify a sharp increase in the persistence of circulation types in the 1980s. More remarkable is the finding that only about half of the variations in European climate since 1780 can be accounted for by change in the frequency of circulation types. The rest is due to within-type changes in the climatic characteristics. For temperatures in January, for example, they show that frequency changes dominated during 1860–1910 and within-type changes during 1910–1980.

7.4 Hemispheric classifications

There are two classifications of hemispheric atmospheric circulation, both developed in the former Soviet Union. One was developed at the Arctic and Antarctic Research Institute, St. Petersburg by G. Ya. Vangengeim in the 1940s–1950s, later modified by A. A. Girs. The other was developed in the 1940s–1960s by B. L. Dzerdzeevski at the Institute of Geography, Moscow.

The main patterns in the Vangengeim–Girs classification, illustrated by Koznuchowsky and Marciniak, are: westerly (W), easterly (E) and meridional (C) in the latitude belt 35–80° N. The patterns describe the distribution of surface cyclones and anticyclones and the major long-wave ridge and trough locations, which vary seasonally. The W category has zonal movements of small-amplitude waves, with nine subtypes according to the latitude of the subtropical anticyclone cells. The C category, with seven subtypes, has large amplitude stationary waves. The subpolar lows are shallow, there is a well-developed high, and the subtropical anticyclone cells are split and displaced poleward. The E category (ten subtypes) is similar to C, but the troughs are in different locations. The subpolar lows are well developed, the Siberian high is weaker, the Azores and Pacific anticyclones are displaced westward and there are stationary highs over Europe and western North America. For 1900–1957 the frequencies of the major types were: W type 26.5 percent; E type 44.4 percent; and C type 29.1 percent.

Meridional (C) circulation dominated in 1890–1920 and 1950–1980. The combined, "zonal" (W + E) circulation epochs dominated in 1920–1950 and 1980–1990. The current "latitudinal" (W) epoch of 1970–1990s is not yet completed. "Zonal" epochs correspond to the periods of global warming and the meridional ones correspond to the periods of global cooling (see Figure 7.2).

B. Dzerdzeevski identified elementary circulation mechanisms (ECMs) based on the hemispheric flow structure over several days. Mid-tropospheric tracks of cyclones and anticyclones indicate the main steering currents. Special attention is given to polar intrusions and blocking in six hemispheric sectors of 50–60° longitude. The four main patterns are:

1 A zonal ring of cyclone tracks in high latitudes; 2–3 breakthroughs of mid-latitude cyclones.
2 A single polar intrusion interrupts the zonality; 1–3 breakthroughs of mid-latitude cyclones.
3 Northerly meridional motion with 2–4 polar intrusions.
4 Southerly meridional motion; no polar intrusions; 1–3 breakthroughs of mid-latitude cyclones.

There are a total of 41 sub-types.

Monthly Northern Hemisphere surface pressure patterns were analyzed by Bartzokas and Metaxas for January–February and July–August 1890–1989 using rotated empirical orthogonal functions to derive the major circulation types. In winter, pattern 1 (27–30 percent of the variance) refers mainly to Europe and Asia, with a strong zonal circulation over the northeast Atlantic and northern Europe; pattern 2 has an anticyclone over eastern Greenland and blocking over northern Europe (21–30 percent of the variance); and pattern 3 features southerly flow over Europe and cold air masses over southern Russia and Siberia (13–17 percent of the variance). In high summer the two main patterns are a polar low pressure (42–45 percent of the variance) and a polar high pressure (42–44 percent of the variance). For summers prior to the 1930s, pattern 2 dominated; since then it has been pattern 1. This shift is correlated with a decrease in summer rainfall over northwestern Europe. There was no trend in winter circulations.

Figure 7.2
Relations between the
detrended global
temperature anomaly
(dt, blue) and the
zonal atmospheric
circulation index
(ACI, red) of
Vangengeim–Girs
(above) and with the
circulation index
shifted four years later
(below) (source:
Klyashtorin 2001).

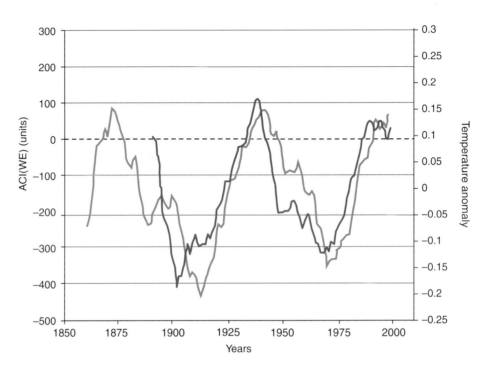

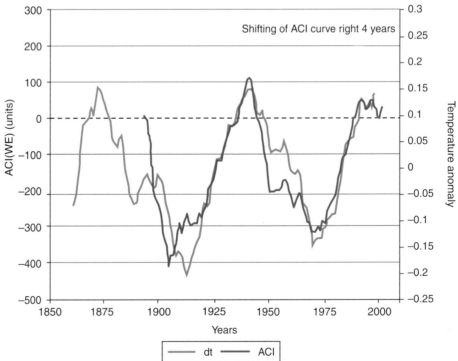

Figure 7.2
Relations between the detrended global temperature anomaly (dt, blue) and the zonal atmospheric circulation index (ACI, red) of Vangengeim–Girs (above) and with the circulation index shifted four years later (below) (source: Klyashtorin 2001).

7.5 Modern applications of synoptic climatology

The traditional uses of synoptic classifications have been illustrated above. More recently, several new directions have emerged. One is the application of synoptic typing to assess general circulation model (GCM) outputs (see Chapter 11). This was first used by Crane and Barry in 1988 in a study of observed and GCM fields for the Arctic. The MSL pressure fields were typed using a method of pattern matching developed by W. Kirchhofer, and also by an EOF method. The simulated patterns appear quite realistic in both frequency and geographical characteristics, although the model data show greater extremes and more closed cells. Subsequently there have been numerous studies of GCM simulations for past conditions and future projections, as discussed by Sheridan and Lee. In general, the observed and GCM patterns match better in winter than in summer. Also, interannual frequency variations are underestimated by GCMs. Multi-model ensembles as well as individual models have been used in the last decade and sometimes ensembles produce better agreement with observations and sometimes not.

Another area of application of airflow pattern or weather types is in assessing teleconnections patterns such as the NAO, PNA, and PDO. For example, Hewitson and Crane demonstrate that the frequency of SOM nodes, derived from eastern North America sea-level pressure data, is correlated with the NAO index. Johnson *et al.* use a SOM approach to examine the differences in the NAO-circulation pattern relationship over the Northern Hemisphere and analyze changes in this relationship during 1978–2005. Several papers have applied synoptic typing to determine the structure of a given teleconnection, such as Cassou *et al.* for the NAO.

SUMMARY

Synoptic climatology is the study of relationships between atmospheric circulation and local or regional climatic conditions. Skilled analysts in the 1930s–1970s developed circulation type classifications subjectively; subsequently objective numerical methods were employed using digital pressure and geopotential height data. These include correlation approaches, empirical orthogonal functions with clustering, and self-organizing maps. Among the best-known and most widely used classifications are those of H. H. Lamb for the British Isles and the *Grosswetter* of Hess and Brezowsky for Europe. Both have been made into objective schemes and kept updated. An air-mass-based approach (spatial synoptic classification) has been developed for local analysis of individual station conditions. Vangengeim–Girs and Dzerdzeevski developed classifications of Northern Hemisphere circulation in the former Soviet Union.

Since 1780, only half of the variations in European climate can be accounted for by changes in circulation-type frequency; the remainder is due to within-type changes. Zonal epochs in the Vangengeim–Girs Northern Hemisphere classification correspond to warm intervals, and meridional ones to cold intervals.

Recent work has explored the use of synoptic typing in evaluating general circulation models against observations and in the analysis of teleconnection patterns.

QUESTIONS

1 Using daily weather maps for your region, attempt to develop a synoptic classification of about ten circulation types and classify the maps for January, April, July, and October for a selected year. Examine the changes in type frequency with season and also analyze the temperature conditions with the major types at a local weather station.

2 Using either the objective Lamb catalog (available at the Climatic Research Unit web site), or the spatial synoptic classification (available at the Kent State University web site), examine the local conditions at any selected weather station in two contrasting winter months.

3 Consider the advantages and limitations of regional-scale and continental-scale synoptic classifications.

4 Describe the three weaknesses of synoptic classifications.

5 How are reanalysis data developed and used in climate studies?

6 Compare air mass and regional airflow classifications.

7 Write a critique of hemispheric classifications.

8 Compare the Vangengeim–Girs and Dzerdzeevski hemispheric classifications.

9 Using the referenced papers in Section 7.5 (Cassou *et al.* 2004; Crane and Barry 1988; Hewitson and Crane 2002; Johnson *et al.* 2008; Kirchhofer 1973; Sheridan and Lee 2010, 2012) examine modern applications of synoptic climatology and comment on its usefulness.

Land and sea effects

The coast at Monterey Bay National Marine Sanctuary, California.

CLIMATIC CONDITIONS are determined to first order by the nature of the underlying surface – whether it is land or ocean – and to second order by the presence of mountain ranges. The principal difference between land and sea lies in the penetration of solar radiation. On land almost all the incoming energy is absorbed at the surface, whereas radiation penetrates several meters into the ocean. In addition the water is in constant motion, mixing the absorbed energy. Consequently, there is almost no diurnal cycle in sea surface temperatures (SSTs), in great contrast to the case of the land surface. Mountain ranges affect weather and climate on local to global scales. Airflow is modified, clouds and precipitation are redistributed, and temperatures are changed both vertically and horizontally. We shall examine global climate through these lenses, starting with the world's oceans.

8.1 Oceans

Globally, 71 percent of the Earth is covered by water, hence oceanic climates dominate the globe. The Southern Hemisphere is 81 percent water and the Northern Hemisphere 61 percent. Unlike the freshwater found in rivers, lakes, groundwater, and precipitation, the oceans are saline waters. Typical ocean salinities are between 32 and 37 parts per thousand. The surface layer is well mixed and warmer due to wave action; below this is a thermocline (vertical temperature gradient) to the deep waters that are cold. Abyssal temperatures average around 1.5–2 °C. The ocean has a mean depth of about 4000 m; the Mariana Trench in the western Pacific reaches 11 km (see Box 8A.1).

The factors that differentiate climate in different parts of the oceans are: latitude, warm and cold currents, and global wind belts. The last of these was treated in Chapter 5. Here we examine ocean currents.

The density of seawater at the surface is between 1022 and 1028 kg m^{-3}, compared with 1000 kg m^{-3} for freshwater. At 0 °C it is 1028 kg m^{-3}, while at 26 °C it is 1023 kg m^{-3}. In polar oceans the density is mainly determined by salinity, while in tropical waters the density is mainly affected by temperature. The distribution of temperature and salinity in the ocean surface waters is more or less zonal. The saltiest waters are in the Subtropics, where there is high

Box 8A.1 **Ocean characteristics**

A vertical thermohaline circulation arises as a result of temperature and salinity contrasts that lead to density contrasts in the ocean. Cold and salty water is denser than warm and freshwater, so it tends to sink beneath the latter. These differences are the main driver of the "conveyor belt" that forms the global ocean circulation (Figure 8.1). The Gulf Stream–North Atlantic Current flows northeastward, cooling as it moves northward and eventually sinking in high latitudes, forming North Atlantic Deep Water. This dense water then drains into the ocean basin and the bulk of it upwells in the Southern Ocean. These components form part of the *Atlantic Meridional Overturning Circulation* (AMOC) or global conveyor belt. Measuring the strength of the AMOC is a challenge, especially in the North Atlantic, where the northern portions are fragmented by the land masses, and most of the sinking takes place in mesoscale eddies. Measurements at 26° N show considerable interannual variability, with values of the AMOC falling from an average of 17 Sv to zero in winters 2009–2010 and 2010–2011, associated with a negative Arctic Oscillation (see Section 5.4) (Sv, for sverdrup, is equal to 1 000 000 m^3 s^{-1}). The water, instead of going northward, turned eastward into the gyre. An analogous event occurred in the winters of 1968–1969 and 1969–1970; the recurrence in a second winter involves the re-emergence in early winter of temperature anomalies preserved in the seasonal thermocline.

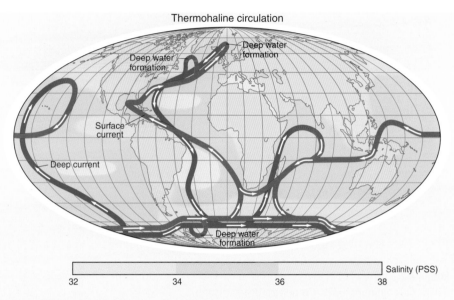

Figure 8.1 The path of the thermohaline circulation/Ocean Conveyor Belt. Blue paths represent deep-water currents, while red paths represent surface currents (source: Wikipedia, http://en.wikipedia.org/wiki/Thermohaline_circulation).

evaporation. The tropical oceans are less salty due to river runoff, as is also the case in the Arctic Ocean due to the northward-flowing rivers in Asia and North America. Also, precipitation in the Intertropical Convergence Zone (ITCZ) tends to lower equatorial salinity. The principal contributor to ocean salinity is sodium chloride (NaCl). The average ocean temperature is only 3.8 °C, whereas the average ocean surface temperature is 16.1 °C and the average land surface temperature is 8.5 °C. Water movement in the deep ocean is extremely slow, less than 0.1 m s^{-1} and so ocean circulations may take 1000–3000 years.

Ocean temperature and salinity have been measured in all the world's oceans since 2007 by 3200 free-drifting Argo floats (small, drifting robotic probes). Prior to their introduction, such data were mostly collected along ship tracks where engine intake readings were taken. The floats move at a depth of around 2000 m, and every nine days they profile to the surface and transmit the data via satellite. The record is not yet long enough to determine trends.

(a) Northern Hemisphere currents

In both the North Atlantic and the North Pacific the southwesterly atmospheric circulation gives rise to northeastward-flowing currents, respectively the Gulf

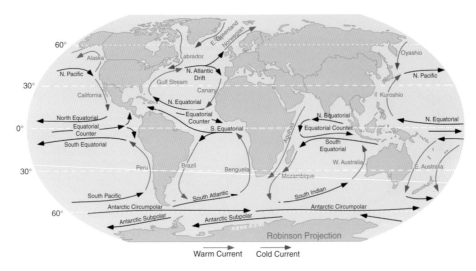

Figure 8.2 The world's ocean currents (source: Wikipedia, http://upload.wikimedia.org/wikipedia/commons/0/06/Corrientes-oceanicas.gif).

Box 8A.2 **The Gulf Stream**

The earliest report on the existence of a transatlantic current was made by Martin Frobisher, who recorded the presence of driftwood from fir trees during his expedition to Greenland and Baffin Island in 1576. Early scientific work on the Gulf Stream was carried out by Benjamin Franklin, who on three Atlantic crossings between 1775 and 1786 observed air and water temperatures and published a map of the Gulf Stream–North Atlantic Current. Between 1808 and 1852 hundreds of bottle tracks were recorded. Northward and southward shifts of warm water off Newfoundland were noted seasonally and on multidecadal time scales, the latter matching southward shifts with intervals of more extensive sea ice off Iceland, according to Stolle and Bryson.

Stream–North Atlantic Current and the *Kuroshio* (Figure 8.2). The *Gulf Stream* originates off Florida and flows northeastward along the eastern coastlines of the United States and Newfoundland (see Box 8A.2 and Figure 8.3) before crossing the North Atlantic as the North Atlantic Current. At about 40° N, 30° W it splits in two, with the northern branch flowing toward northern Europe and the Norwegian Sea, and the southern branch re-circulating off the Canary Islands and West Africa. The Gulf Stream is about 100 km wide and 800–1200 m deep. Off Florida it transports about 30 Sv and off Newfoundland about 150 Sv. The Gulf Stream is wind-driven, and at the surface reaches a velocity of ~2.5 m s^{-1}. The water temperature is about 21 °C off Florida, decreasing by

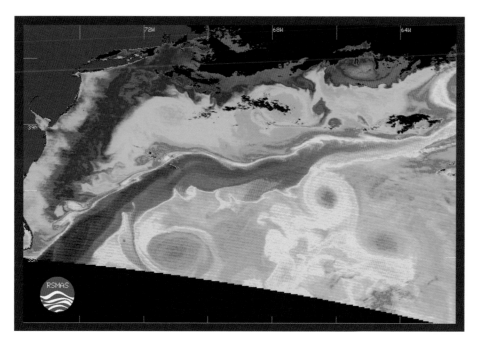

Figure 8.3 The sea-surface temperature in the Gulf Stream on May 8, 2000 as seen by MODIS. The sea-surface temperatures range from about 8 ° (violet) to 20 °C (red). The image was created at the University of Miami using the 11 and 12 μm bands, by Bob Evans, Peter Minnett, and co-workers (source: Visible Earth NASA, http://eoimages.gsfc.nasa.gov/images/imagerecords/54000/54734/gulf_stream_modis_lrg.gif).

evaporative cooling to 8 °C off Scotland. The North Atlantic Current represents a continuation of the Gulf Stream that is also maintained by the ocean thermohaline circulation (see Box 8A.1). It is partly responsible for the mild winter climate of northwest Europe and for keeping the west coast of Norway free of sea ice in winter. In the Labrador Sea cold Arctic water flows southward, for much of the year carrying sea ice and Greenland icebergs from Baffin Bay with it. Water temperatures are − 1 °C in winter and about 5 °C in summer. The Labrador Current has a significant cooling effect on the Atlantic provinces of Canada and coastal New England. It encounters the Gulf Stream at the Grand Banks southeast of Newfoundland, and this gives rise to frequent advection fogs as warm air flows over the cold water. The cooling of the surface air leads to it becoming saturated, giving rise to condensation as fog droplets. Fog is most common between November and March, but the Grand Banks average over 200 days per year.

There is also a cold East Greenland Current flowing southward from Fram Strait along the east coast of Greenland, which carries Arctic sea ice far south. The current turns after rounding Cape Farewell at the southern tip of Greenland and flows northward along the west coast of Greenland.

In the subtropical North Atlantic there is a massive clockwise oceanic gyre. This is formed by the Gulf Stream and North Atlantic Current on the west and north, respectively, by the Canary Current on the east, and by the westward-flowing North Atlantic Equatorial Current on the south that is driven by the easterly trade winds. These four currents comprise a gyre that encloses the Sargasso Sea, which spans 70–40° W, 25–35° N.

In the North Pacific, the *Kuroshio* begins off the east coast of Taiwan and flows northeastward past Japan, where it merges into the easterly drift of the North Pacific Current. It is analogous to the Gulf Stream–North Atlantic Current, and the cold southward-flowing California Current is analogous to the Canary Current in the eastern North Atlantic. Off northeastern Asia, the cold subarctic waters of the Oyashio Current flow southward, analogous to the Labrador Current.

In the northern Indian Ocean the currents reverse with the seasons in response to the northeast winter monsoon and the southwest summer monsoon. Off Somalia in summer there is a very strong current from the southwest, driven by a low-level atmospheric jet stream over East Africa.

(b) The Arctic Ocean

The *Arctic Ocean* is a unique water body that covers a little over 14 million km^2, is almost surrounded by land masses – North America, Greenland, and Eurasia – and is totally covered by sea ice in winter. It is the shallowest of the world's oceans and receives about 4200 km^3 of freshwater discharge annually, about 60 percent from northern Asia. As a result, its surface salinity is only about 33 percent and its surface temperature averages about −1.7 °C, close to the freezing point for sea-water. Below about 50 m depth, the temperature and salinity both increase. The warmer, more saline Atlantic enters the Arctic Ocean via the Norwegian and Barents seas at intermediate depths (200–900 m). In the Beaufort Sea, north of Alaska and northwest Canada, there is a large clockwise gyre that is driven by a quasi-permanent high-pressure center in the winter half-year. From Siberia to the Fram Strait, off east Greenland, there is a Transpolar Drift Stream that transports sea ice out of the Arctic Ocean into the Greenland Sea.

The Arctic Ocean begins to freeze in October and *sea ice* becomes continuous over the Arctic Ocean and the waters of the Canadian Arctic Archipelago. It extends into Baffin Bay, the Labrador Sea, Hudson Bay, the Barents Sea, the Bering Sea, and the Sea of Okhotsk, with a total area of around 15 million km^2 in late winter. There is also sea ice in Bohai Bay off eastern China at latitudes 38–40° N!

Young ice is highly saline, because sea salts are trapped in the ice, but during the winter brine drains out into the ocean, making ice that is low in salinity. Along the coasts of the Laptev and East Siberian seas and in the Canadian Arctic Archipelago, winds may drive ice away from the land, forming areas of open water and thin young ice known as *polynyas* (a Russian term). These polynyas are major sources of heat fluxes to the atmosphere and are also vital habitat for many marine mammals. In the cold Arctic winter, polynyas tend to rapidly freeze over. The winter ice has a snow cover that is about 30–40 cm deep. In summer, the snow melts, forming melt water ponds on the ice, which in turn melts and thins.

From 1979 to the mid 1990s the summer minimum in September averaged around 6.5 million km^2, but there was a long-term downward trend that accelerated in the 2000s. The winter trend in ice area was -3 percent per decade and the summer trend -12 percent per decade up to 2011. Between 2007 and 2011, the summer minimum averaged only 4–5 million km^2. In September 2012 a record minimum of 3.4 million km^2 was recorded (Figure 8.4). Moreover, the ice thinned dramatically between the 1970s and 2000s, from 3.1 m to about 1.5 m, and the proportion of old, multiyear ice dropped sharply. In the 1950s–1980s, there was typically 40 percent first-year ice and 60 percent older ice. In the late 1990s–2000s these percentages reversed as much old ice was exported out of the Arctic Ocean. Two other factors that account for the thinning and ice retreat are incursions of warm southerly airflow

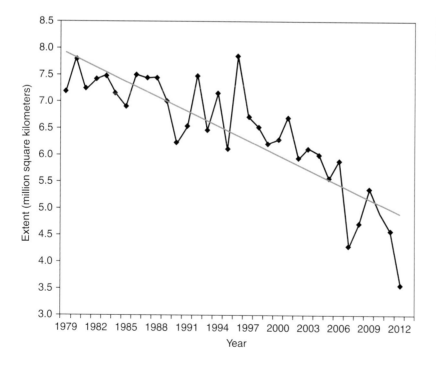

Figure 8.4
The trend of Arctic sea ice extent in September 1979–2012 (source: NSIDC).

over the Arctic basin and the advection of warm water from the North Pacific Ocean via the Bering Strait into the Chukchi and Beaufort seas.

(c) Southern Hemisphere currents

In the Southern Hemisphere there are broad anticlockwise gyres in the three oceans that feature cold northward-flowing currents off southwest Africa (the Benguela Current), off western South America (the Humboldt or Peru), and off western Australia (the Leeuwin) and westward-flowing equatorial currents in each ocean. On the western boundaries of the oceans there are warm southward-flowing currents – the Algulhas (or Mozambique) off east Africa, the Brazil Current off Brazil and Argentina, and the East Australia Current. To the south there is a broad, eastward-flowing *Antarctic Circumpolar Current* (ACC) that is continuous around the hemisphere. It is up to 2000 km wide and 2000–4000 m deep, with speeds of 20–40 cm s^{-1}. It transports an enormous amount of water, of the order of 125 Sv. The ACC keeps warm ocean waters away from Antarctica, enabling the continent to maintain its enormous ice sheets. SSTs decrease from about 5 °C at 55° S, in the Drake Passage, south of South America, to 0 °C at 60° S. SSTs in the eastern South Pacific display cold water related to the Peru Current and cold upwelling caused by the offshore winds along the tropical west coast of South America. The cold water turns westward in a narrow current along the Equator. In the western equatorial Pacific off New Guinea there is a warm pool with surface temperatures above 29 °C. The warm pool is an important locus of convection and plays a role in tropical cyclone development.

(d) The Southern Ocean

The waters in the *Southern Ocean* off Antarctica have extensive sea ice in winter, covering about 20 million km^2 at maximum in September. The ice averages about 0.9 m in thickness with a deep snow cover. In the growth phase the weight of snow depresses the ice floes, allowing seawater to flood the ice and soak the snow. This then freezes into snow ice, a process that is absent in the Arctic. By late summer (March), the sea ice has greatly retreated, with about 3 million km^2 remaining, mainly in the Weddell Sea. Close to the coast of Antarctica there is extensive landfast ice (attached to the land) that can grow thermodynamically by heat conductive losses to 2 m in thickness during the winter season.

The southern oceans are colder than their northern counterparts due to the presence of Antarctica and the Southern Ocean, which is much colder in the austral summer than the Arctic in the boreal summer. As a consequence the locations of the thermal equator and the ITCZ between the two trade wind systems are displaced northward from the geographical Equator, to around 5° N.

The westerly winds in the Southern Ocean are uninterrupted by any land masses, so high velocities are the norm. This has given rise to the designation of the Roaring Forties between latitudes 40° and 50° S; even stronger winds occur in the Furious Fifties (50° S) and the Screaming Sixties (60° S).

8.2 Air–sea interaction

Air–sea interactions involve the exchange of heat, water, momentum, and gases between the atmosphere and ocean. A described in Section 2.1g, heat transfers include both sensible and latent heat. Transfers of water involve precipitation, evaporation, and horizontal atmospheric moisture transport. Mid-latitude oceans take up carbon dioxide, accounting for about one-quarter of the CO_2 emitted by human activity, and this contributes to ocean acidification. The spatial scale of air–sea interaction ranges from meters to thousands of kilometer teleconnections. Time scales range from seconds to years and centuries.

The ocean currents largely determine the pattern of SSTs, which in turn affect the overlying air masses. From the Subtropics, tropical maritime air moves poleward over cooler sea surfaces, becoming more stable, while polar maritime air moves equatorward over warmer sea surfaces, becoming more unstable and moist. In the northwestern Atlantic, in January, the air is $> 6\,°C$ colder than the sea surface.

Over the oceans, the annual and diurnal variations of temperature are much less than those over land areas in the same latitudes. This is due to the high heat capacity of water, which means it warms up slowly and cools slowly. The heat capacity of water is 4.1855 kJ $(kg\ K)^{-1}$, whereas that for soils is about four times lower – of the order of 1 kJ $(kg\ K)^{-1}$. As a consequence of this contrast, air temperatures over land lag solar radiation by about two hours on a daily basis, peaking around 2 p.m., and by a month on a seasonal basis; the solar minimum is at the winter solstice (21 December), whereas the average minimum temperature for the Northern Hemisphere occurs around 19 January.

In contrast, SSTs have negligible diurnal change and lag solar radiation by two to three months seasonally.

On a larger scale, the oceanic current systems like the Gulf Stream, the Kuroshio, and the Antarctic Circumpolar Current support atmospheric baroclinic (frontal) zones and help steer extratropical cyclones along them. This idea was explored extensively by Jerome Namias in the 1970s (see Box 8B.1). Sea surface temperatures in autumn and winter in the area south of Newfoundland have been shown

Box 8B.1 Jerome Namias

Jerome Namias was born in Bridgeport, Connecticut in 1910. He trained at the University of Michigan and in the late 1930s did research on the Dust Bowl at the Massachusetts Institute of Technology. From 1941 to 1971 he was Chief of the Extended Forecast Division of the US Weather Bureau, now the National Weather Service, and in the 1940s developed the five-day forecast. In the 1960s he developed monthly and seasonal forecasts and analyzed the interactions between the atmosphere and ocean, in particular the role of Pacific SSTs and El Niño on the climate of North America. In 1971 he joined the Scripps Institution in La Jolla, California and established the first Experimental Climate Research Center. He died in 1997.

to influence the climate of the northeastern Atlantic. Positive departures of 1–2 °C lead to pressure values 3–4 mb below average in the northeastern Atlantic in the following month.

In the Southern Hemisphere there are three significant zones of airflow convergence, giving rise to convective cloud bands that extend southeastwards from heat sources in Brazil: the South Atlantic convergence zone; over Indonesia and the Pacific warm pool (the South Pacific convergence zone [SPCZ]); and in central Africa (the South Indian Ocean convergence zone), all in the austral summer. The SPCZ, which has been most studied, involves the Southern Hemisphere ITCZ in its tropical section west of 180° longitude, and wave disturbances on the South Pacific polar front along its extratropical section. The cloud band extends to 25° S, 150° W according to Vincent. The other two do not have extratropical elements and are only linked to the continental heat sources. Their formation and maintenance are poorly understood. They weaken and disappear in winter as the subtropical high-pressure cells shift equatorward.

Over subpolar oceans in both hemispheres there is a distinctive type of weather system known as the *polar low*. These have a diameter of only 500–1000 km and a lifetime of a couple of days. They typically form near the sea ice margin, where there are steep temperature gradients, and beneath an upper cold low. They spin-up in 12–18 hours. One type has a spiral pattern of cumulonimbus clouds around an eye, analogous to the tropical cyclone. The other is a cloud system shaped like a large comma (known as a comma cloud). The former tends to occur in polar air, the latter nearer to the polar front. They bring severe weather, with snow and hail, and winds of 20–30 m s^{-1}. They dissipate rapidly when they move onshore. They occur during November–March in the Northern Hemisphere and year-round in the Southern Ocean.

Marine stratocumulus cloud decks commonly occur in the subtropical latitudes off the west coasts of the major continents related to the cold ocean currents noted above. The cloud decks over the eastern North and South Pacific and the eastern South Atlantic are well-known examples because they often affect the west coast strip of their respective continents. The eastern North Atlantic typically has less extensive cloud cover, as water temperatures there are somewhat higher.

Downstream of the zones of stratiform cloud, to the west, skies are nearly clear, with scattered cumulus cloud formed in the trade wind flows, typically aligned in "cloud streets" parallel to the wind direction. Subsiding air in the subtropical anticyclones gives rise to a trade wind inversion of temperature and this limits the vertical development of the clouds. Further downwind over the central oceans the subsidence weakens and cloud tops penetrate the inversion, gradually spreading moisture upward. Also, cloud clusters and traveling waves form along the ITCZ.

In the Caribbean, traveling waves, known as *easterly waves*, are common during April/May–October/November. They were first described by Herbert Riehl in the 1940s (see Box 8B.2). There are about two systems per week traveling westward at 20–35 km h^{-1}. The waves are 2000–2500 km long and consist of an open waveform extending northward from the equatorial trough of low pressure. Ahead of the wave there is divergence at low levels and subsidence giving little cloud, whereas behind the wave there is low-level convergence that gives rise to convection and rainfall. Eventually, over the warm waters of the western oceans, some of these disturbances may develop into tropical storms and hurricanes. In a similar fashion there are waves and shallow depressions in the central and western Pacific, some of which develop into tropical storms and typhoons.

Box 8B.2 Herbert Riehl

Herbert Riehl was born in Germany in 1915. In 1933 his mother sent him to England to perfect his English. He emigrated to the United States in 1933. During the 1940s he worked at the Institute of Tropical Meteorology in Puerto Rico, where he provided a description of the easterly wave. In 1954 he wrote the first classic text on "Tropical Meteorology" and lectured on it at the University of Chicago. He carried out studies of tropical convection with Dr. Joanne Simpson and they developed the concept of "hot towers" (deep cumulonimbi) that transport heat into the upper tropical troposphere. He became known as "the father of tropical meteorology." He moved to Fort Collins, Colorado in 1962 to begin the Department of Atmospheric Sciences at Colorado State University, from where he worked on the meteorology of Venezuela. He died in 1997.

Areas of deep convective activity in the western equatorial Pacific Ocean exhibit a quasi-periodic behavior, recurring over about 7–21-day intervals. The systems drift slowly eastward from 140° E to 180° E.

(a) Tropical cyclones

Tropical cyclones form from pre-existing tropical waves or depressions that move out of West Africa, or redevelop in the eastern Pacific after crossing Central America. Those in the northwest Pacific develop from tropical waves and depressions in the central Pacific. The waves in West Africa have a wavelength of about 3500 km and occur about every 4–5 days from June to early October. The cyclones are known as hurricanes in the western North Atlantic, typhoons in the northwest Pacific and cyclones in the southwestern Indian Ocean. Their development takes place over warm surface waters, where temperatures are $> 27\,°C$ (see Section 3.4 and Figure 8.5); most form over waters between 28 and 30 °C. As a consequence of the absence of such high temperatures in the South Atlantic, only one tropical cyclone has ever been recorded there.

Northern Hemisphere cyclones typically move westward within the tropical easterlies and then recurve northward and northeastward over eastern North America and eastern Asia as they are steered by the upper southwesterly winds. However, some Atlantic systems enter the Caribbean and Gulf of Mexico. A few of these cross Central America and redevelop in the northeast Pacific off Mexico. Cyclones in the southwest Pacific curve southwestward toward northeastern Australia, while those in the southwest Indian Ocean curve southwestward toward Madagascar. Systems in the Bay of Bengal, which originate from westward-moving depressions in the tropical easterlies, impact Bangladesh and eastern India. Occasionally, a system forms in the Arabian Sea. Table 8.1 summarizes the mean annual frequency of tropical storms and cyclones in the major ocean basins.

8.3 Land

Land occupies 39 percent of the Northern Hemisphere and 19 percent of the Southern Hemisphere. The average elevation is only 840 m. The climates of land areas are determined by many factors: latitude, distance from the ocean, mountain ranges, vegetation and snow cover, and synoptic weather patterns.

Latitude has major effects on incoming solar radiation amounts (Figure 2.3), which decrease in winter to zero at the Arctic Circle. In summer, radiation receipts are less variable with latitude owing to the increase in *day length*, with latitude

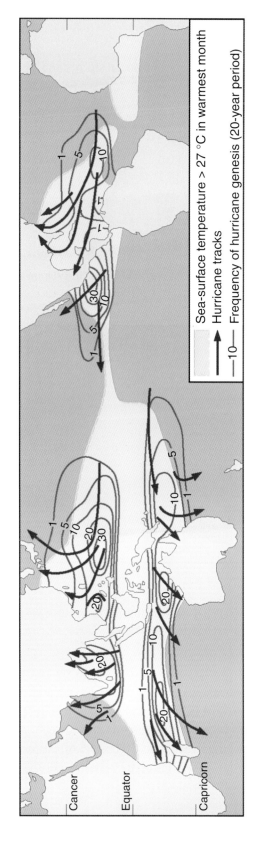

Figure 8.5 Frequency of tropical cyclones, tracks, and sea-surface temperatures >27 °C in the warmest month (source: Barry and Chorley 2010).

Table 8.1 Annual average number of tropical storms and tropical cyclones in different ocean basins ranked by frequency

Ocean basin	Tropical storms	Tropical cyclones
Northwest Pacific	26.7	16.9
South Indian	20.6	10.3
Northeast Pacific	16.3	9.0
North Atlantic	10.6	5.9
Australia/southwest Pacific	9.0.	4.8
North Indian	5.4	2.2
Total	88.6	49.1

Source: Wikipedia. http://en.wikipedia.org/wiki/Tropical-cyclone

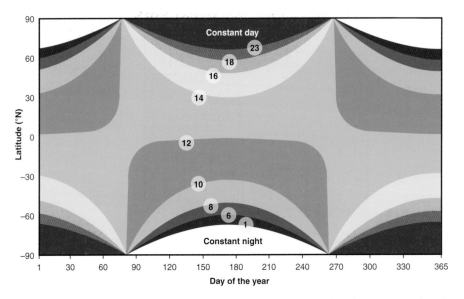

Figure 8.6 The latitudinal variation of day-length over the annual cycle (source: Wikipedia, http://upload.wikimedia.org/wikipedia/commons/b/b9/Hours_of_daylight_vs_latitude_vs_day_of_year.png).

offsetting the low solar elevation angle. Annual temperatures decrease with latitude; they vary diurnally in low latitudes, and annually in high latitudes as a result of the diurnal and annual cycles of solar radiation and daylight. At the Equator the day length is close to 12 hours year round (Figure 8.6). At the Poles the sun is above/below the horizon for six months of the year and there is almost no diurnal temperature variation.

Distance from the ocean in the direction of the prevailing winds exerts a major influence on the annual temperature range. Near the coast this range is small, but it increases rapidly as one moves inland. The annual temperature range (July–January monthly means) is used as a measure of *continentality*, which is largest in the continental interiors. The highest mean annual temperature range is an astonishing 60 °C in northeast Siberia and > 40 °C in the Yukon of northwest Canada. In contrast, it is only 10–15 °C in maritime coastal regions in middle latitudes. There are similar small ranges in tropical latitudes due to the constant, almost-overhead sun.

With increasing distance from the upwind ocean, precipitation amounts decrease and cloud bases rise in response to the drying out of the air as a result of rainfall. In Europe annual precipitation amounts decrease from 500–750 mm in north-central France to 300–500 mm in north-central Poland and in Ukraine, for example. The drop off in rainfall amounts moving eastward in mid-latitudes is greatly strengthened by the presence of mountain ranges along the west coasts of North and South America, northwest Europe and New Zealand. Mountain summits may receive around four times the amount that falls upwind and on the lee side there is a rain shadow where totals may decrease to only one-quarter to one-third of the upwind amount. In western Scotland at sea level annual totals are around 1250 mm and this decreases to < 600 mm in the eastern part of the country. In the South Island of New Zealand, annual totals exceed 10 m on the west slopes of the Southern Alps and decline to below 700 mm east of the mountains.

Vegetation cover affects climate through its albedo, evapotranspiration, and surface roughness. The albedo ranges from 0.10 over forests to 0.25 over grassland, hence causing greater absorbed solar radiation over forests. The evaporative response is shown to be very important in heat-wave conditions. Tower measurements in Europe show that, initially, surface sensible heating is 2–4 times higher over forest than over grassland according to Teuling *et al*. Over grass, heating is suppressed by increased evaporation in response to increased solar radiation and temperature. Ultimately, however, this process increases soil moisture depletion in the grassland and induces a critical shift in the regional climate system that leads to increased heating. Because forests evaporate less water in the initial phase of a heat wave, they can sustain a constant level of evapotranspiration over a longer period of time than grasslands.

Surface roughness is a function of canopy height and plays a role in evaporation losses through its effects on the vertical gradient of wind speed. In general, greater roughness values lead to increased evaporation. Values of surface roughness length range from ~0.04 m for long grass to about 2.0 m for tropical forests. Its value is approximately one-tenth the height of surface roughness elements. It represents the height at which the wind speed above the ground theoretically decreases to zero.

Snow cover has major effects on climate, operating in at least four main ways. The first is albedo–temperature feedback. This operates when snow cover shrinks, lowering the albedo from about 0.8–0.9 to about 0.1–0.3. This reduction increases the absorption of incoming solar radiation at the surface by about four times on average. The warming of the surface leads to a further reduction of the snow cover, hence positive feedback. The effect works in reverse with increased snow cover, leading to enhanced cooling and further extension of the snow cover. The second effect involves the insulation of the ground surface by snow cover. This cuts off the transfer of heat and moisture from the underlying surface to the atmosphere (or vice versa), which modifies the thermal regime of the underlying ground and of the overlying atmosphere. Winter air temperatures in the presence of snow cover on the North American prairies are 5–10 °C lower than with bare ground. The third effect is on the hydrological cycle, caused by the temporary storage of water in snow cover. This delays the annual runoff peak. The fourth effect involves the release of latent heat during the phase changes from vapor to liquid (the latent heat of vaporization is 2500 kJ kg^{-1}), and liquid to solid ice (the latent heat of fusion is 333 kJ kg^{-1}), and the corresponding heat inputs during evaporation and melt. Hence, evaporation requires nearly eight times more energy than melt.

Direct sublimation of snow to vapor requires 2833 kJ kg^{-1} (the sum of the latent heat of fusion and of vaporization) and thus is less common than snow melt.

SUMMARY

Oceans occupy 71 percent of the Earth's surface and consequently play a major role in global climate. Their large heat capacity means they heat up and cool down slowly. Ocean currents in the Northern Hemisphere transport heat northeastward in the North Atlantic and North Pacific, keeping western Europe and southern Alaska, respectively, much warmer in winter than the corresponding latitudes in eastern North America and eastern Asia. Cold currents flow southward off east Greenland, Labrador, and northeastern Asia. A major part of the global thermohaline circulation (Atlantic Meridional Overturning Circulation) comprises the northward flow in the North Atlantic, with dense water sinking in high latitudes and returning southward at depth, much of it upwelling in the Southern Ocean off Antarctica.

The Arctic Ocean and adjacent seas are covered with 2–3 m thick sea ice in winter. Ice extends to 40° N off China. The ice shrinks greatly in summer and since 2005 has reached the lowest levels on record in September. The ice has also thinned considerably and much less of it is now multiyear ice than was the case from 1979 to 2000. Large amounts of multiyear ice have been exported from Fram Strait via the Trans-Polar Drift Stream. In the Southern Hemisphere, north-flowing cold currents affect the western coasts. Off Peru the cold water turns westward in a narrow band along the Equator. The broad Antarctic Circumpolar Current flows continuously around the Southern Ocean. In September, sea ice about 0.9 m thick off Antarctica covers 20 million km^2, which shrinks to only 3 million km^2 in March.

In the Southern Hemisphere there are three atmospheric convergence zones that extend southeastward from continental heat lows. The South Pacific Convergence Zone is unique in having both tropical and extratropical components. In the eastern subtropical oceans a major feature is the presence of extensive marine stratocumulus. Downstream in the trade winds, cloud cover eventually breaks up. In the Caribbean and in the western North Pacific there are wave systems traveling westward between April and November. About 10 percent develop into tropical storms or cyclones.

In the subpolar oceans of both hemispheres, mesoscale polar lows traveling eastward are an important element of weather and climate, bringing severe weather conditions.

Tropical storms and cyclones form from pre-existing waves or depressions in low latitudes and travel slowly westwards. They form in all tropical oceans except the South Atlantic. They are most frequent in the northwestern Pacific, the South Indian Ocean and the northeastern Pacific. Cyclones have inward-spiraling cloud bands, very strong winds, and an internal eye, 50–100 km wide, where winds are light. Many recurve over the western oceans into the mid-latitude westerlies and may reform as an extratropical low-pressure system.

Over land, the annual temperature range increases greatly away from windward coasts. Maximum continentality is in northeast Siberia and the Yukon. Precipitation amounts decrease inland away from windward coasts, especially in the lee of mountain ranges where there is a rain shadow. Mountain summits and windward slopes may receive four times the annual totals upwind.

Vegetation cover affects climate via its albedo and surface roughness. Forests can sustain high evapotranspiration rates much longer than grassland. The climatic influence of snow cover operates via albedo–temperature feedback, surface insulation, water storage, and phase change effects on latent heat.

QUESTIONS

1 Compare the annual and diurnal cycles of solar radiation and temperature at the Equator and Poles.
2 Explain how ocean currents affect coastal temperatures around the North Atlantic and the North Pacific.
3 What factors determine ocean salinity?
4 What factors explain the distribution of tropical cyclones?
5 Rank the following surface types according to their albedo: agricultural land, ocean, desert, boreal forest.
6 How does vegetation cover affect climate on large and small scales?
7 Explain the concept of continentality.
8 Discuss, with examples, the role of snow cover in climate on large and small scales.
9 Explain the role of heat capacity in the daily and annual temperature regimes of land and ocean.

Climatic types on land

The Southwestern desert, United States, with saguaro cactus.

IN THIS chapter we describe the major climatic types found in continental areas, especially in terms of their temperature and precipitation regimes. The distribution of these types, in the context of the Köppen classification of climates, is illustrated in Figure 1.3. We begin by discussing approaches to climatic classification. The main classifications are based on the seasonal characteristics of temperature and moisture, but more specialized ones are based on thermal comfort or agricultural conditions. There are also classifications based on the causes of climatic patterns.

We begin by considering some of the most spatially extensive types of climate – deserts, monsoons, and high plateaus. In each case, regional differences are discussed. Then we turn to wet lowlands and move poleward through the different latitude zones to the polar regions and ice sheets of Greenland and Antarctica.

9.1 Classifying climates

There are many approaches to classifying climates. The earliest dates back to the Greeks, who recognized the changes of temperature with latitude. They identified torrid, temperate, and frigid zones. The primary variables that characterize a region are generally its temperature, rainfall, and, in particular, their seasonal characteristics. Essentially, temperature values can be divided into about five classes and precipitation distinguished according to its seasonal distribution (year round, summer maximum, winter maximum). This is the basis of W. Köppen's classification referred to in Section 1.4. H. Kraus and A. Alkhalal provide budgets of Rn, LE and H over the annual cycle for ten of Köppen's types (see Section 2.1). Another approach is to consider the moisture balance (precipitation minus evaporation) and its seasonal distribution, as proposed by C. W. Thornthwaite in the United States in 1948. Numerous authors have developed variants of these schemes – A. Miller, W. Gorcynski, and G. Trewartha, for example. Others, such as K. Buettner and W. Terjung, developed physiological classifications of thermal comfort. J. Papadakis developed an agriculturally based classification. Genetic classifications based on air mass frequencies and airstream boundaries have been proposed by W. Wendland and R. Bryson. A very thorough annotated bibliography on climatic classifications has been compiled by R. F. Strauss.

It must be recognized that all classifications are arbitrary and the boundaries between categories will sometimes be indistinct. However, where there is a topographic feature, such as a mountain range or a coastline, the boundary will be sharp.

9.2 Major climatic types on land

(a) Deserts

Deserts (BW climates according to Köppen) occupy approximately 25 percent of the global land area based on the Köppen classification. Hence, they are major elements of global climate. They are regions that are almost devoid of vegetation, with very low rainfall, typically < 250 mm per year, but their temperature conditions vary greatly from one to another. Their surfaces may be sandy with giant dunes, stony gravel, or bare rock pavements. Dry river channels, which are occasionally subject to flash floods, traverse them.

Deserts are formed in several ways. Major ones develop in the Subtropics, where persistent anticyclones and associated subsiding air create clear skies and low relative humidity. The hot deserts of North Africa – the 9.1 million km^2

Sahara – and of Arabia in the Northern Hemisphere, and Australia in the Southern Hemisphere, are prime examples. The high-pressure centers that are responsible are apparent in Figure 3.2. A second type occurs within interior basins that are remote from moisture sources and surrounded by mountain ranges. The mid-latitude Gobi (44° N, bounded by the Altai Mountains and the Tibetan Plateau) and Taklamakan (41° N, bordered by the Kun Lun, the Pamir, and the Tien Shan ranges) in northwest China are examples. A third type is coastal desert in the Tropics and Subtropics, where cool offshore currents add to the effects of subsiding air in the subtropical high-pressure belts. Fog and low stratus frequently form in the cool, moist ocean air and are drawn inland by sea breezes. The Namib Desert in southwest Africa and the Atacama Desert in northern Chile are examples. Coastal areas of Namibia have >100 days of fog annually. The associated fog drip on cactus vegetation is a crucial moisture source. The Atacama is the driest desert in the world because moisture is blocked by the Andes to the east and by the Chilean Coast Range to the west. Some locations appear never to have received any rainfall. A fourth type of desert is the cold polar desert attributable to low atmospheric humidity and infrequent weather systems that result in low precipitation. In polar deserts the ground is usually frozen and the permafrost (permanently frozen ground) may be hundreds of meters thick.

Temperature conditions in deserts vary widely according to their latitude and continentality. The Sahara is very hot in summer (Figure 9.1a) but relatively cold in winter, with occasional frost and snow on the 3000 m mountains (the Hoggar and Tibesti) within it. The Gobi and Taklamakan deserts are extremely cold in winter, with occasional snow, and hot in summer. Polar deserts have annual precipitation of less than 250 mm, falling as rain in summer and snow in autumn, with a mean temperature during the warmest month below 10 °C. Winter temperatures average −40 to −50 °C. Polar deserts are found in the Canadian Arctic Archipelago, northern Greenland, and the Siberian Arctic islands, as well as in the Dry Valleys of Antarctica, west of McMurdo Sound. In the last area, mountains block potential moisture sources and downslope katabatic winds from the Antarctic Plateau sublimate snow and ice.

Annual precipitation thresholds for desert climates depend on the seasonality of the precipitation. Winter precipitation is subject to much less evaporation than that falling in the summer. Relative humidity also displays strong seasonality. In Riyadh, Saudi Arabia, values in the warmest month (35 °C) range from 30 percent at 1400 hours to 45 percent at 0700 hours. In the coolest month (14 °C), however, corresponding values range from 45 percent to 70 percent.

Conditions in the lower atmosphere over the deserts vary greatly with location and season, as illustrated in Figure 9.1 for Hotan and Tamanrasset. The deserts in interior basins experience strong temperature inversion conditions in winter,

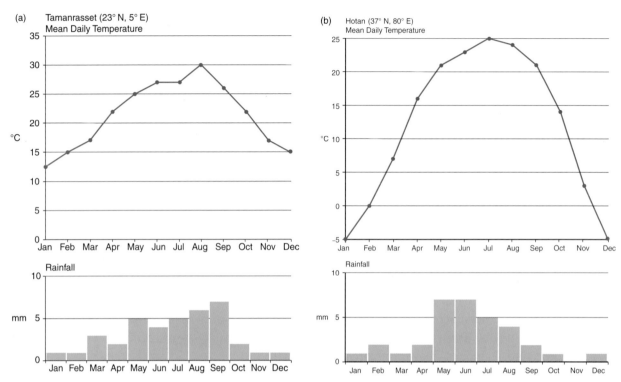

Figure 9.1 The annual cycle of temperature and rainfall at (a) Tamanrasset in the Sahara and (b) Hotan in the Taklamakan.

producing especially low temperatures. Temperatures in the Taklamakan desert in the Tarim Basin fall as low as –20 °C, for example. Conversely, in the Sahara in summer there may be highly unstable air in the boundary layer due to surface heating. The instability can generate dry convection, giving rise to dust devils. These form when a plume of hot surface air rises rapidly through cooler air above. They develop when the vertical temperature gradient at the surface exceeds 10 °C/100 m (a super-adiabatic lapse rate). If conditions are just right, the air may begin to rotate. The column of hot air is stretched vertically, causing intensification of the spinning effect by the conservation of angular momentum. At the surface, hot air flows inward, intensifying the vortex. They typically raise a plume of dust.

With occasional strong winds and thunderstorm gust fronts passing over desert areas, fine sand and dust may be lifted into the atmosphere, generating a *dust storm*. In North Africa these are termed haboobs. They can be up to 100 km wide and 1–2 km deep, traveling at 35–50 km hr^{-1}. Hence, deserts are sources of dust aerosol. Saharan dust is transported over the North Atlantic and sometimes across the Mediterranean into Europe, where it can be observed in red rains. Dust from the Gobi desert blows across northern China and into the North

Pacific. In southern Arizona there are 7–9 dust storms annually, with visibility reduced to below a kilometer in half of them.

The major deserts exert an important influence on the climate of the adjacent areas. Between December and mid-March the northeasterly harmattan blows from the Sahara over West Africa. It brings very dry, cool, and dusty conditions. In November and March, hot, dry, continental tropical air from the Sahara may be drawn northward into southern Europe (the scirocco) by depressions moving eastward over the Mediterranean Sea. Wind speeds may exceed 25 m s^{-1}. Temperatures are high and the air is dry over coastal North Africa, but cool, damp air behind the depressions affects southern Europe. In South Australia, a hot northerly wind from the desert often precedes the eastward passage of a frontal system in summer. It brings hot, dusty conditions to the south of the country and is analogous to the scirocco of North Africa.

Desert margins in many parts of the world are showing signs of *desertification* (desert expansion due to land degradation) as a result of both climate change and human activities. It is estimated that a total global area of 6–12 million km^2, is affected by desertification. Degradation of vegetation cover, especially savanna grassland, as a result of overgrazing has been identified as a major factor. However, the issue is complex. In the Sahel of West Africa a prolonged decrease in rainfall since the 1970s has been a driving factor. The solution is to reduce livestock numbers, but this may be difficult with rising populations.

(b) Monsoons

The major *monsoon* regions of the Tropics extend from Australia/Indonesia in the east, through Southeast Asia and the Indian subcontinent, and across Africa to West Africa. There is a separate region in northwest Mexico and the southwest United States, and another in central South America.

The term *monsoon* is derived from the Arabic word "mawsim," meaning season. It refers to the major seasonal wind reversal over southern Asia from northeasterly in winter to southwesterly in summer in response to the different patterns of land/sea heating and temperature contrast. Basically, air flows from the cooler ocean to the heated land in summer and in the opposite direction in winter. However, this is a very simplistic view. There is a heat low over northern India in spring with air flowing toward it, but this is well before the onset of the southwest monsoon and its heavy rains.

To understand the South Asian monsoon we need to examine the upper air circulation over the region. In winter, the upper-level westerlies are split by the Tibetan Plateau, and the subtropical jet stream flows eastward south of the Himalaya, while the Polar Front jet stream is well to the north over Siberia

(Figure 9.2a). Air subsides beneath the exit of the subtropical jet stream due to high-level convergence as the air slows down, and this sinking air gives rise to the low-level northeasterly winds that flow out of India over the Indian Ocean.

In May the subtropical jet stream begins to shift northward and in June the heating of the Tibetan Plateau (which averages 4500 m elevation, or ~ 600 mb) creates a shallow heat low and an upper-level anticyclone. The anticyclone gives rise to upper-level easterly winds over India and forms the Tropical Easterly Jet stream at around 150 mb (13.5 km) at about 15° N. At low levels, warm, moist southwesterly flow over the Indian Ocean pushes northeastwards toward India (Figure 9.2b). However, the monsoon onset begins over Burma and Assam because the rain-bearing systems are monsoon depressions that travel to the west and northwest, steered by the high-level easterlies.

Some 4–6 of these depressions form over the Bay of Bengal and affect India each monsoon season. The southwesterlies bring rain to the Western Ghats along the west coast of India, but it is the monsoon depressions and occasional tropical cyclones from the Bay of Bengal that spread rainfall to the north and west, so that by July they reach southern Pakistan and the western Himalaya. Some moisture penetrates into southern Tibet, but there the rainfall tends to occur in convective showers at night. Bangalore in south-central India receives about 100 mm per month during May–August (Figure 9.3), but then 300 mm in September as the monsoon retreats, because the Bay of Bengal depressions are then moving across India at lower latitudes.

The monsoon has irregular active and break intervals over India. The active phases of convection occur when the monsoon trough is in a northerly location and breaks when it moves to the south. The active periods typically occur at 15–20-day intervals and last about 10–15 days. The frequency of active periods during the summer season determines how much precipitation falls during the monsoon. There is large interannual variability in monsoon rains – in both the timing and amount. For all-India, the monsoon provides 78 percent of the total average rainfall. The mean amount for 1871–1990 was 852 mm, with a range from 70 to 120 percent about the mean according to Parthasaraty *et al*. Totals were low from 1901–1930 and 1961–1990, and high from 1881–1900 and 1931–1960. New work shows a progressive decrease in all-India rainfall since the mid twentieth century, from 871 mm to 811 mm in 2001–2010, as a result of warming in the western Pacific warm pool and an eastward shift of the monsoon rain.

During September–October the upper easterlies weaken and disappear as the upper anticyclone over Tibet decays, setting the stage for the return of the subtropical jet stream over northern India. In winter, low-pressure systems, steered by the subtropical westerly jet stream from the Mediterranean and Middle East, enter Pakistan, Kashmir, and northwest India, bringing rain and snow to the Karakorum Mountains.

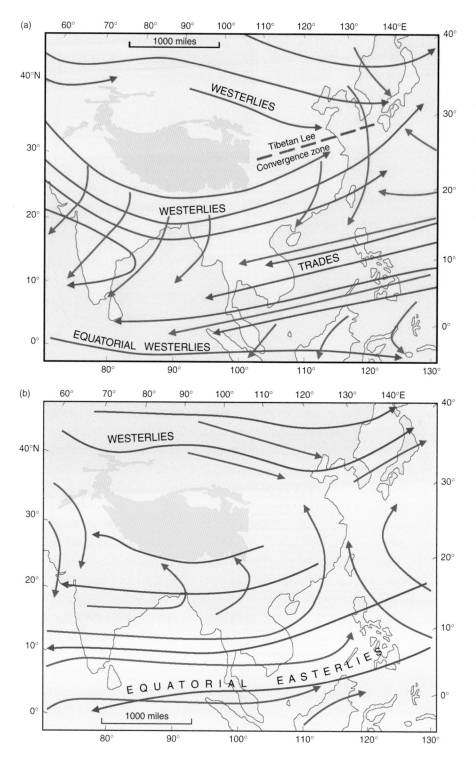

Figure 9.2
The circulation over southern Asia in (a) winter (b) summer. The flow lines show upper-level circulation (brown) and low-level circulation (blue) (source: Barry and Chorley 2010).

Figure 9.3
Annual cycle of temperature
and rainfall on a hythergraph
plot for Bangalore, south-
central India (blue)
and Darwin, Australia (red).

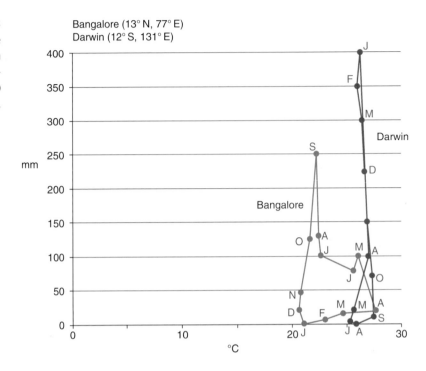

Figure 9.3 Annual cycle of temperature and rainfall on a hythergraph plot for Bangalore, south-central India (blue) and Darwin, Australia (red).

Over China in winter there are frequent cold air outbreaks from the northwest related to the cold winter anticyclone over eastern Siberia. In spring this weakens and disappears. The summer monsoon arrives in late May in the south and in late July in the far northwest of China. The moisture source is southerly flow from Indonesia and the South China Sea, not the southwesterlies over South Asia. The spring–summer Maiyu (plum rains) front appears to involve both the monsoon trough and the East Asian–West Pacific Polar Front. Rains in China are linked to weak disturbances that move eastward along the Yangtze and occasional cold fronts from the northwest.

Over West Africa the monsoon southwesterlies advance from the coast of the Gulf of Guinea in February to about 13° N in May, but then in mid-June rains occur widely from 9–13° N. The extreme northern limit of surface monsoon air (the Intertropical Discontinuity or Front) is at about 20° N; however, a sufficient depth of moisture for precipitation reaches only to about 13° N. Aloft between about 10° and 17° N there is a weak African Easterly Jet stream (AEJ) at about 3 km, overlain by the stronger Tropical Easterly Jet stream at 15–20 km, which is a continuation of the system over India.

The AEJ is a result of the temperature gradient between the Sahara and the Gulf of Guinea. It shifts north–south seasonally with the advance and retreat of the monsoon. As is the case over India, waves move westward, steered by the tropical easterlies. Some of these waves may later develop into tropical storms in the western Atlantic. Associated with the jet streams are deep cumulonimbus clouds developed on westward-moving north–south-oriented disturbance lines. To the south there is a zone of disturbances that originate over East and Central Africa. These give rise to deep stratiform cloud with prolonged light rainfall. At the coast there is a narrow zone of frictional convergence that gives light rainfall within the monsoon air that is overlain by subsiding easterlies.

In late December (austral summer) a thermal low develops over northern Australia with low-level westerlies, overlain in the upper troposphere by easterly winds, analogous to the summer pattern over India. Beginning over Borneo in early November, northwesterly flow over Indonesia and New Guinea in December penetrates to northern Australia. The monsoon season at Darwin averages about 75 days from 28 December to 13 March, based on the winds between the surface and 500 mb. Darwin shows a pronounced wet three months in January–March. However, the duration of the monsoon season is highly variable. Also, as in India, there are active periods of 4–14 days and breaks of 20–40 days. Much of the precipitation is associated with monsoon depressions, of which there are typically five per season.

In the American southwest and northwest Mexico there is also a summer monsoon (the Mexican or North American monsoon). The driving factors appear to be the heating of the Colorado Plateau area in June–July that leads to the development of an upper-level anticyclone (analogous to Tibet and the Altiplano; see Section 9.2c). Moist southerly flow affects northwest Mexico in mid-June and reaches southeast Arizona and southwest New Mexico around 1 July. Over 70 percent of annual precipitation in northwestern Mexico falls in June, July, and August (JJA), and over 50 percent in southeast Arizona and southwest New Mexico. Monsoon failures are implicated in major droughts in the southwestern United States. Surges of moisture from the Gulf of California affect lower elevations, and southeasterly flow from the Gulf of Mexico at 700 mb affects the higher elevations further north. The rain generally falls in brief, but intense, afternoon–evening thunderstorms that form into clusters, triggered by surface heating.

The monsoon flow continues northward, intermittently affecting Colorado and Utah. Interactions of very moist air with the Rocky Mountains have resulted in some dramatic rainstorms and floods. The Big Thompson flood near Estes Park, Colorado, on 31 July 1976 was the result of >300 mm rain falling in four hours, leading to the death of 144 people who were trapped in the narrow canyon. Moist tropical airflow led to a record 435 mm rainfall during 9–16 September 2013 (rated a 1000-year event) in Boulder, Colorado, with disastrous flooding in canyons along the Front Range.

Meteorologists Vera *et al.* and Marengo *et al.* now consider that central South America experiences a monsoon system, although it straddles the Equator rather than being in the Tropics, and its seasonal temperature variations are less. Nevertheless, central South America receives over half its annual rainfall in the monsoon season. In September the onset of the wet season starts in the equatorial Amazon and then spreads to the east and southeast during October. By late November deep convection covers most of central South America from the Equator to 20° S, but is absent over the eastern Amazon basin and northeast Brazil. During late November to late February the main convective activity is centered over central Brazil and linked with a band of cloudiness and precipitation extending southeastward from southern Amazonia toward southeastern Brazil and the surrounding Atlantic Ocean, known as the South Atlantic Convergence Zone (SACZ). The strong convective heating over the Amazon sets up a Chaco low east of the Andes over northern Argentina (Chaco is a lowland in interior south-central South America). During March–May, precipitation amounts decrease and rainfall shifts north toward the Equator. Further discussion of the Amazon is offered in Section 9.2d.

The area of global monsoons, as defined by the Global Precipitation Climatology Project (GPCP), is ~80 million km^2 (almost 16 percent of the global surface), and over the period 1979–2008 the area increased by 0.22 million km^2 yr^{-1}. This increase in area has given rise to an increase in global monsoon precipitation, but also to a decrease in monsoon intensity (defined as the global monsoon precipitation amount per unit area). The expansion of the monsoon area is perhaps caused by a widening of the tropical circulation in the poleward branch of the Hadley cell.

(c) High plateaus

There are two major high plateaus: Tibet (Qinghai-Xizang), which covers 2.5 million km^2 and averages ~4500 m in elevation; and the much smaller Altiplano of Bolivia in South America.

The *Tibetan Plateau*, called "the roof of the world," and also known as the "Third Pole" (see Box 9B.1) stretches approximately 1000 km north to south and 2500 km east to west. It is bordered on the south by the Himalayan mountain ranges and it encompasses several mountain ranges (notably the Kun Lun and Tanggula) within itself. In the 3000 km long Kun Lun in the north, which rises to over 7000 m, there is a glacier area of 12 200 km^2; in the moist Nyeqentangula (or Nyainqentanglha) east of Lhasa there is 10 700 km^2 of ice, and in the central Tanggula Mountains ice covers 2200 km^2. Temperatures are low in winter, although there is generally only light snowfall due to the dry air. In the higher western parts average winter (summer) temperature is − 20 to − 25 °C (5 °C) and in the lower eastern areas they are, respectively, − 10 °C (15 °C). In the north, the high plateau of southern Qinghai and northern Tibet is underlain by continuous permafrost (permanently frozen ground) 100–300 m thick, and this becomes discontinuous and sporadic in the south. The railroad from Golmud to Lhasa,

Box 9B.1 **The "Third Pole"**

The term "Third Pole" is currently used to describe a unique high mountain region centered on the Tibetan Plateau that stores more ice and snow than anywhere else in the world outside the polar regions. The Third Pole contains the world's highest mountains, including all 14 peaks above 8000 m, and is the source of ten major rivers.

With an average elevation of over 4000 m, and an area of more than 4.3 million km^2, the Third Pole is characterized by complex interactions of atmospheric, cryospheric, hydrological, geological, and environmental processes that have a significant effect on the Earth's biodiversity and climate and hydrological cycles. These processes are critical for the people of surrounding regions of Afghanistan, Bangladesh, Bhutan, China, India, Kazakhstan, Kyrgyzstan, Myanmar (Burma), Nepal, Pakistan, Tajikistan, and Uzbekistan. The melt water from 12 000 km^3 of glaciers of the Third Pole supplies permanent flow to most of Asia's major river systems.

There are significant climate impacts and feedbacks associated with this unique region. The Third Pole plays a prominent role in the evolution of the Asian monsoon system, which is critical for the precipitation pattern in the region, and snowfall amounts on the plateau can both weaken or prolong the summer monsoon system in the region.

Conditions in the Third Pole affect the atmospheric circulation patterns in Eurasia and thus significantly influence the climate system in the Northern Hemisphere. The fluctuations of climate have strong effects on the glaciers and lakes of the Third Pole, but also on water supply and social stability for the wider region. As glaciers retreat, water volume and flow are likely to decline and could become seasonal rather than year-round in some of the drier regions. Additionally, the observed thawing of permafrost will result in the release of greenhouse gases, such as methane and CO_2 to the atmosphere, contributing to global climate change.

built in 2006, crosses 550 km underlain by permafrost that necessitated specially engineered embankments to prevent the ground from thawing.

Annual precipitation totals in Tibet decrease northward and northwestwards from >2000 mm in the southeast to 50–100 mm in the northwest. The Asian monsoon influences the southern margins of the plateau in July (Lhasa, Figure 9.4), starting in the southeast and moving westward. Precipitation is convective, from large cumulonimbus cells, sometimes with hail. The snowline reaches 6000 m on the north side of the Himalaya and decreases northward at about 100 m per degree of latitude.

The high plateau of the *Altiplano* (18° S) has an average altitude of 3750 m and covers about 100 000 km^2, with the Andes mountains to the west and dissected mountains to the east and north. It has a dry steppe climate. Analogous to Tibet, an upper-level anticyclone (the Bolivian high) forms over the Altiplano in austral summer (January). Rainfall over the Altiplano is <700 mm and declines from

Figure 9.4
The annual cycle of
temperature and rainfall shown
by hythergraphs for Lhasa, Tibet
(red), and Juliaca, Peru (blue).

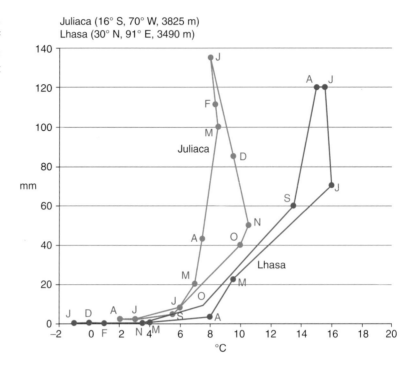

north to south. It is mainly concentrated in the austral summer months (Juliaca,
Figure 9.4), especially in its southwestern part, where over 70 percent of the
precipitation occurs in December–February according to Garreaud *et al.* It arises
from deep, moist convection where the moisture is transported westward from
Amazonia, located to the east of the Altiplano. The rainfall occurs in wet episodes
lasting about a week, separated by dry spells of similar length. Southward displacement of the Bolivian high gives rise to stronger easterly flow aloft. However,
much of the moisture is derived from regional slope circulations along the eastern
Andes that draw in moisture from the boundary layer over the adjacent lowlands.
Interannual variability of rainfall in the region is linked to ENSO, with cold La Niña
phases generally being wet. However, the 1988/1989 dry La Niña and the 1972/
1973 wet El Niño seasons in the Altiplano were exceptions. In these years the
spatial pattern of zonal wind anomalies led to these different responses. In May–
October, mean westerly flow dominates the Altiplano region, resulting in dry
conditions. The highest mean temperature of ∼20 °C is in spring (November),
with a value of 13 °C in winter. Minimum temperatures are around 1–5 °C in
summer and –11 °C in winter, with a winter diurnal range of 25–40 °C as a result of
the pronounced radiational heating and cooling.

A prominent feature of the Altiplano is Lake Titicaca, with a surface area of 58 000 km². The influence of such an impressive body of water is felt by a reduction in diurnal temperature range and the net effect is to increase the mean annual temperatures by 1–2 °C within a few tens of kilometers of the lake margin above those in the surrounding region. This sets up a local climatic belt around the lake. The lake, with water warmer than the surrounding air once the sun sets, also influences precipitation, which is greatest at the lake's center. Annual precipitation totals over the lake are increased by about 15 percent above those in the surrounding area. Mean annual precipitation over the lake is 880 mm versus an average of 760 mm in the surroundings. Lake Titicaca, because of its area and volume and its situation at high altitude within the Tropics, remains a hydrological site unique in the world.

(d) Wet lowlands

The major *wet lowlands* of the world – Af climates according to Köppen – are in the great equatorial river basins of the Amazon in South America and the Congo in Africa.

The Amazon River basin covers about 7 million km². Its thermal regime is distinguished by its constancy. Temperatures vary diurnally but show almost no seasonal change. Mean monthly temperature at Manaus ranges between 26 and 29 °C, but the daily range is about 8 °C. Annual precipitation at Manaus is 2280 mm, with a maximum in January–April and a minimum in July–September (Figure 9.5a). Over the basin annual totals range from 3500 mm in the west between 4° N and 4° S, to 1500 mm in the east. Typical rainfall amounts of 2500 mm are associated with evapotranspiration amounts of 1200–1500 mm annually. About 75 percent of the moisture for precipitation is advected from the ocean, while at least 25 percent is recycled evaporation from the land. Deforestation is having important impacts on the hydrological cycle. Removal of forest leads to a reduction of surface roughness and an increase in albedo that lower evapotranspiration and therefore precipitation, while the albedo increase also results in less moisture flux convergence in the atmosphere, further reducing precipitation. The diurnal characteristics of tropical rainfall are summarized in Box 9A.1.

In the dry season there are easterlies from the tropical Atlantic that have a shallow, relatively cool, humid layer overlain by drier, warmer air. In central and southern Amazonia, there is a continental heat low from October to April, with substantial precipitation. In November cold fronts may affect the southern margins of the basin. In northern Amazonia the rainy season is from May to September, associated with low-level convergence in the equatorial low-pressure trough and instability lines. The ITCZ is only evident along the coast, where it migrates southward to about 5° S from December to April and then northward to

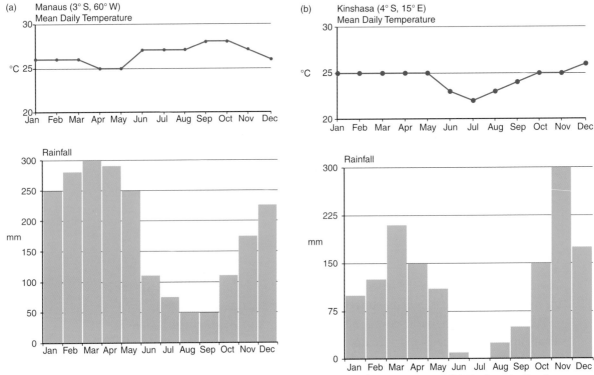

Figure 9.5 The annual cycle of temperature and rainfall at (a) Manaus in the Amazon and (b) Kinshasa in the Congo.

10° N in September to October. In the east, a significant source of rainfall is instability lines formed by coastal convergence in the trade winds, or convergence between nocturnal land breezes and the trade winds. Most of these instability lines die out as they travel about 100 km inland, but some persist for 500 km and a few may reach the eastern Andes. Extratropical disturbances that cross the Andes from the South Pacific affect the mid-latitudes and Subtropics; between 20° S and the Equator there are about two such systems per month throughout the year. These contribute to the rainfall in southern, southwestern, and west central Brazil according to Satamurty et al.

In the austral winter, cold air surges (friagems) from the south affect southern Brazil, when a Pacific low-pressure system enters Chile and Argentina. Low temperatures in June–July are a major hazard for Brazil's coffee production and so raise coffee prices.

During the wet season of January–February 1999 in southwest Rondônia, Silva Dias et al. found that there were alternating periods of westerly and easterly flow. The former gave rise to stratiform cloud with generally light rain and the latter to convective cloud with heavy rainfalls. These regimes appear to be linked to the

Box 9A.1 **Diurnal character of tropical rainfall**

The classical picture of tropical rainfall is that it has a distinct diurnal cycle. Cumulus cloud builds up as the surface is heated during the day, leading to the formation of cumulonimbus (see Chapter 2 plate) in the afternoon, downpours of rain, and thunderstorms. Rainfall rates diminish in the evening hours, reaching a minimum in the early morning hours. Where mesoscale convective systems are important, however, land areas have a peak rainfall from late evening to midnight as a result of their longer life cycle compared with individual thunderstorms. Over tropical oceans there is only a weak diurnal signature. There tends to be an early morning peak, which is associated with the occurrence of mesoscale convective systems (see Section 3.5).

20–50-day Madden–Julian Oscillations in the equatorial region (see Section 6.11). An active South Atlantic Convergence Zone (SACZ) leads to a westerly wind regime over Amazonia, whereas breaks in the SACZ lead to the easterly regime.

The *Congo River basin* covers 4 million km^2 astride the Equator. The climate of Congo is equatorial, with the rainy season related to the passage of the ITCZ. In January, the ITCZ is oriented north–south from 15° S to 5° N over the central Congo basin, between westerly flow from the South Atlantic and easterly flow from the Indian Ocean. To the west along about 15° S there is the Inter-Ocean Convergence Zone (IOCZ), also known as the Zaire Air Boundary (ZAB), between northwesterly flow from the South Atlantic and south easterlies from the Indian Ocean. In July the IOCZ is oriented southwest–northeast from 15° S to the Equator, as shown by van Heerden and Taljaard. The disturbances associated with the convergence zones are either heat lows or waves and lows traveling westward in the tropical easterlies. The lows extend up to about 600 mb, overlain by an anticyclone. Another source of precipitation is mesoscale line squalls that move westward.

In the north the rainy season is from March to June. The moisture is brought at upper levels by the northerly component of the AEJ. As the ITCZ moves southward, the maximum precipitation is in July and August; it then shifts into central Congo in September and October, when the moisture comes via low-level advection from the Atlantic Ocean. Between November and February the southern parts receive maximum precipitation. Thereafter, the ITCZ moves northward again, crossing central Congo in March and April, so this zone has two rainfall maxima (Kinshasa, Figure 9.5b). Precipitation totals are around 1800–2000 mm, less than in Amazonia. This is in part because the East African highlands block moisture from the Indian Ocean according to Trewartha. In the regions with a dry season, annual totals are of the order of 1500–1600 mm. Temperatures and relative humidity ($\sim$80 percent) are constantly high in the central region, with mean temperatures about 27 °C and rarely dropping below 20 °C.

(e) Tropical and subtropical steppe

Tropical steppes (BS climates) occur on the equatorial margin of tropical deserts, as in the Sahel grasslands of Africa. Subtropical steppe is found on the poleward margin of tropical deserts. Annual rainfall in the Sahel is 100–500 mm, falling in summer from instability lines that move westward (see Section 9.2b). Mean monthly temperatures range from about 23 °C to 34 °C.

Like their mid-latitude counterparts, they periodically experience protracted droughts. The Sahel was particularly affected in the 1970s–1980s, although the late twentieth century remained much drier than the first 60 years of the century. There had been similar events in the 1740s–1750s and 1820s–1830s. The primary cause is thought to be a decrease in sea-surface temperatures (SSTs) in the tropical South Atlantic and it has been suggested that the Atlantic Multidecadal Oscillation (AMO) in ocean temperatures plays a role. The AMO, whose full cycle spans an interval of $\sim$70 years, was negative in the 1970s–1980s. However, studies show that SSTs across the tropical oceans also play a role in Sahel droughts. Relations have been demonstrated with both El Niño and Indian Ocean SSTs. These remote teleconnections must involve convective build-up, perhaps via the MJO and its effects on upper-level flow in the Walker circulations.

(f) Humid subtropical

The *humid subtropical climate* (Cf or Cw according to Köppen) is most developed in the lowlands of the southeastern United States and southeastern China, but also occurs in southeastern South America and coastal southeastern Australia. There are some notable differences between the regions, particularly in eastern Asia.

The region of the southeast United States is dominated by tropical maritime air that flows northeastward from the Gulf of Mexico. Summers are long, hot, and humid, with daily average temperatures above 25 °C. There are only a few days with freezing temperatures in winter and snowfalls are rare. Inland in spring and summer the area may be affected by violent thunderstorms and tornadoes that move east or northeastward. The peak frequency occurs in Oklahoma in late spring. A second maximum is found along the Gulf Coast in late autumn. Per 26 000 km^2 there are eight tornadoes per year in Oklahoma and nine in Florida. They commonly form where cold, dry polar air from the northwest moves over warm, moist, southerly tropical air. This leads to the formation of a supercell thunderstorm, where high instability and a strong moisture source favor tornadoes. This gives rise to a southwest–northeast-oriented belt that is referred to by weather forecasters as "tornado alley." The mechanism that initiates rotation is still unclear. They occur most often in early evening. The scale of tornado intensity, originally developed by T. Fujita in 1971, now ranges between F0 with winds of 105–137 km hr^{-1}, and F5 with winds over 322 km hr^{-1}. On average, there are about 800 tornadoes per year in the United States, resulting in 80 deaths and 1500 injuries.

In southeast China the winters are cooler than in the southeast United States as a result of cold air outbreaks from eastern Siberia. In summer the region is affected by the East Asia–West Pacific monsoon. Much of the water vapor in summer originates over the Indian Ocean, but the west Pacific high exerts strong control on the monsoon over southeast China. Annual rainfall totals exceed 1500 mm near the coast and are 1000–1500 mm to the north and west. Shanghai has temperatures averaging 5 °C in January and 28 °C in July, whereas the corresponding figures for Hong Kong are 17 °C and 28 °C.

The coastal regions of both the southeastern United States and southeast China may be affected by tropical cyclones in summer and autumn. These bring high winds and heavy rainfall. In the southeastern United States up to 15 percent of the hurricane season (June–November) rainfall is brought by tropical cyclones. In the southeastern coastal regions of China, tropical cyclones (typhoons) bring more than 500 mm of rain annually, accounting for 20–40 percent of the total annual precipitation.

In southeastern South America the humid subtropical climate includes the Pampas grassland region. There is rainfall throughout the year with a minimum in June. In Buenos Aires, summer temperatures average 25 °C and in July (winter) they average 11 °C. Cold spells may occur in winter with outbreaks of polar air from Antarctica. In July 2007, snow fell in Buenos Aires for the first time since 1918.

(g) Mediterranean

Mediterranean climates (Cs according to Köppen) are found at about 35° latitude in California, southwest Australia, central Chile, and southwestern South Africa, in addition to around the Mediterranean Sea. The classical definition involves a mild winter with precipitation and a dry, hot summer. However, around the Mediterranean Sea itself, rainfall regimes are variable. The classical pattern is common in the central and eastern Mediterranean (Rome, Figure 9.6) but in Spain, southern France, northern Italy, and the northern Balkans there is an autumn maximum or a double peak in autumn and spring.

Figure 9.6
Annual cycles of temperature
and rainfall in a hythergraph
plot for Rome (blue) and
Los Angeles (red).

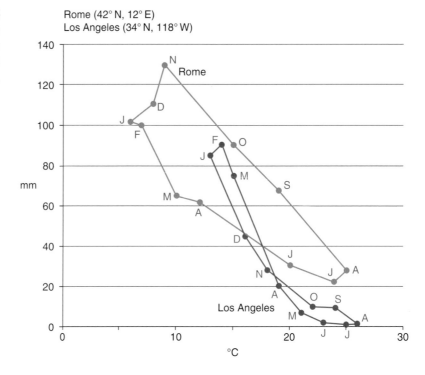

Temperatures are generally moderate year round, although in the more continental eastern Mediterranean there are hot summer days and, in winter, frosts
and snowfall. Overall, the proximity to water moderates the annual temperature
range. The climatic conditions are related to the seasonal shifts of the subtropical
high-pressure belts – poleward in summer and equatorward in winter. In winter
most of the regions are affected by low-pressure systems steered by the subtropical
jet stream. However, in the Mediterranean basin 74 percent of depressions form
over the western Mediterranean Sea in the lee of the Alps associated with northwesterly flow that splits around the Alps. Only 9 percent enter from the Atlantic
and the rest form south of the Atlas Mountains in Morocco. In California the zone
with a Mediterranean-type climate is limited to a narrow coastal strip by the
mountains to the east (Los Angeles, Figure 9.6).

(h) Temperate lowlands

Temperate lowlands are both continental and maritime. Temperate climates
are found in a range of locations between 40° and 60° latitude. They occupy
most of northeastern North America, central and eastern Europe, and European
Russia, and account for about 5 percent of the Earth's land surface. From Europe

Berlin (53° N, 13° E)
Christchurch (43° S, 173° E)

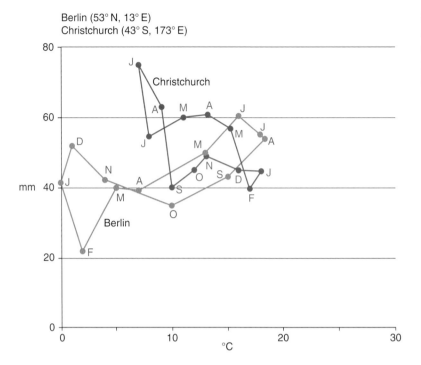

Figure 9.7
Annual cycles of temperature and rainfall in a hythergraph plot for Berlin (blue) and Christchurch (red).

to Russia the continentality increases eastward (Berlin, Figure 9.7). They are also found in New Zealand's South Island, east of the Southern Alps (Christchurch, Figure 9.7), and in southeastern Australia. They include temperate grasslands and deciduous woodlands, but today they are mostly in temperate agriculture.

Temperatures are seldom extreme and all four seasons are well marked. In central England summer temperatures range between 15 and 20 °C, while the cool winters seldom see daytime temperatures drop below zero. Average annual precipitation is in the range 600–1500 mm, usually without any dry season. Winter snowfalls are generally moderate; they occur only on a few days in the maritime locations, but snow may lie on the ground for weeks in the uplands and in the continental portions.

Extratropical depressions and intervening high-pressure ridges regularly affect temperate zones. Hence, there is a high degree of day-to-day variability in weather conditions and long spells of a given weather type are rather uncommon. The prevailing winds are from the west; thus the western edge of temperate continents most commonly experience maritime temperate climate. Here there is a very limited annual temperature range (10–15 °C) and precipitation occurs year round in moderate amounts.

(i) Maritime west coasts

Maritime west coasts are found in northwest Europe, British Columbia, southern Alaska, and also in coastal Chile and western New Zealand. In each region this climatic type is located on the windward slopes of major mountain ranges. In northwest Europe it extends further inland due to the southwest–northeast orientation of the mountains. These locations are in the path of year-round westerly winds and traveling mid-latitude low-pressure systems, giving cloudy, mild, and humid conditions with large day-to-day variations in weather. Exposed coastal and upland locations experience generally high wind speeds and frequent winter gales. Wind gusts in upland Britain occasionally exceed 50–60 m s^{-1}. The annual temperature range is small. In Brittany, northwest France, the mean temperature is 7 °C in January and 16 °C in July. The cold season has the most rainfall. Annual precipitation totals range from ~1000 to 3000 mm (Figure 9.8 for Forks, WA and Bergen, Norway), and on the coastal mountains there may be significant winter snowfall. At 1660 m elevation on Mt. Rainier, WA, the mean annual snowfall totals 16 m, with a record 28 m falling in winter 1971–1972.

Figure 9.8
Annual cycles of temperature and rainfall in a hythergraph plot for Forks, WA (blue) and Bergen, Norway (red).

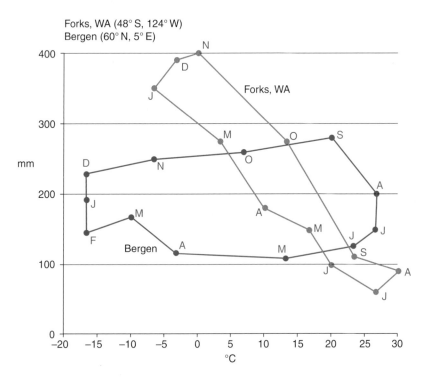

Forks, WA (48° S, 124° W)
Bergen (60° N, 5° E)

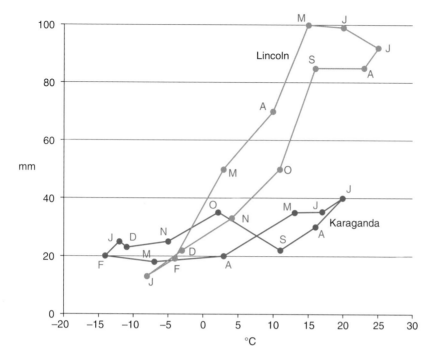

Figure 9.9
The annual cycles of temperature and precipitation plotted as hythergraphs for Lincoln, NE (blue) and Karaganda, Kazakhstan (red).

(j) Mid-latitude steppe and prairie

Steppe is a Russian term for dry grassland that is cold in winter ($-30\,°C$) and hot ($35\,°C$) in summer (BSk in Köppen's classification). It occurs most extensively in southern Siberia, Mongolia, Tibet, and northwest China, but also in the Midwestern United States, where it is called prairie, and in Patagonia. The annual rainfall is between 250 and 600 mm, most falling in summer convective events. Winter snow cover is generally thin, but may cover the ground for weeks to months. Figure 9.9 shows the annual cycles of temperature and precipitation at Lincoln, NE, and Karaganda, Kazakhstan. Steppe areas are typically located in continental interiors remote from moisture sources, or are in the lee of major mountain ranges. They are susceptible to large interannual variability of rainfall and to periodic droughts. The "Dust Bowl" years of the 1930s on the US Great Plains are a good example (Box 9B.2).

(k) Taiga/boreal forest

The *boreal forest* is the largest terrestrial biome, covering most of Canada, southern Alaska, much of Scandinavia, and Russia (where it is known as the taiga). It covers 14.5 percent of the land surface. The climate is subarctic (Df), with a wide

Box 9B.2 **The Dust Bowl**

In the 1930s the Midwest of the United States experienced a severe and protracted drought. For the entire growing season (April through August), the precipitation total during the Dust Bowl years was the least in 160 years, but equivalent to that in the mid nineteenth century in northeast Kansas. This was a period of massive dust storms, which caused major agricultural damage in the American and Canadian prairies, primarily from 1930 to 1936, but in some areas until 1940. These conditions were due to the combination of severe drought and decades of extensive farming without crop rotation. The deep plowing had killed the natural vegetation that normally kept the soil in place and trapped moisture even during dry periods and high winds.

Tons of topsoil were blown off barren fields and carried in storm clouds for hundreds of miles. The driest regions of the Plains – southeastern Colorado, southwest Kansas, and the panhandles of Oklahoma and Texas – became known as the Dust Bowl, but the entire region, and eventually the entire country, was affected. Millions of acres of farmland became useless; soon, hundreds of thousands of people were forced to leave their homes. The Dust Bowl exodus was the largest migration in American history within a short period of time. By 1940, 2.5 million people had moved out of the Plains states, headed primarily for the west coast.

In 1933 President Franklin D. Roosevelt established governmental programs designed to conserve soil on the Great Plains. Additionally, the Federal Surplus Relief Corporation (FSRC) was created to stabilize pricing of agricultural commodities and distribute food to families nationwide. Roosevelt also ordered the Civilian Conservation Corps to plant a huge belt of more than 200 million trees from Canada to Texas to break the wind, hold water in the soil, and hold the soil itself in place. By 1937, education programs were set up to teach farmers about soil conservation and by the following year, the conservation effort had reduced the amount of blowing soil by 65 percent. However, it would be two more years before the drought was over, and farmers could once again grow crops.

annual temperature range and especially severe winters. The lowest temperatures in the Northern Hemisphere are recorded in the river basins of northeastern Siberia. Yakutsk is the coldest city in the world, with a January mean temperature of −40 °C. A Northern Hemisphere record low temperature of −71 °C was established at Omyakon in the Yana valley. Winters last 5–7 months with substantial snow packs, and much of the ground is underlain by permafrost. In central Siberia permafrost depths exceed 1000 m. Summer lasts only two to three months, with mean temperatures above 10 °C. Annual precipitation is relatively low – generally 250–750 mm, with 1000 mm in some areas – primarily falling as rain during the summer months.

In summer, low-pressure systems on the polar front regularly cross the boreal forest zones of Alaska, Canada, northern Europe, and Russia. In winter, however, northwest Canada is frequently dominated by high pressure and in eastern Siberia there is a shallow but intense and persistent cold anticyclone – the Siberian high. Scandinavia and western Siberia experience both traveling lows and highs.

In Canada the Boreal Ecosystem–Atmosphere Study (BOREAS) was carried out during 1994–1996. Results demonstrated the low forest albedo – only 0.08 in summer and even with winter snow cover around 0.25 – and the fact that one-third of the study area had been burned within the last 25 years, as reported by Sellers *et al.* Low rates of evapotranspiration were observed (~ 2 mm d^{-1}), despite the high fraction of lakes and wetland. Sensible heat fluxes exceeded latent heat.

In Eurasia there has been a wide range of studies under the Northern Eurasian Earth Science Partnership Initiative (NEESPI). Bulygina *et al.* demonstrate changes in snow cover during 1966–2007. There has been an increase in mean snow cover depth in Yakutia and in the Far East. The winter-average snow depth has maximum trends in north West Siberia. Mean snow depths for the period with permanent snow cover increased considerably over the four decades, from 10 to 35 cm at Dudinks (69° N, 86° E) and from 30 to 60 cm at Turukhansk (66° N, 88° E), both in Yakutia.

(l) Tundra

Tundra means "treeless" in Lapp (Finland) and this zone (ET according to Köppen) occurs north of the tree line in Arctic Canada, Arctic Siberia, the Siberian islands, coastal Greenland around the ice sheet, and in Svalbard. Alpine tundra is found in high mountain areas worldwide at altitudes that range from about 5000 m in equatorial latitudes, to 6000 m in the Tropics, and around 3000–4000 m in mid-latitudes, declining to near sea level in high latitudes. The vegetation is mainly grasses and low vascular plants, with shrubs occupying sheltered locations in the southern part of the zone. Winters are long and cold, while summers are short, with temperatures around 5–10 °C. In the polar regions the sun is below the horizon for six months of the year. In winter, temperatures are around -25 °C to -30 °C, except when low-pressure systems bring warmer air masses from the North Pacific or North Atlantic.

Figure 9.10 shows mean monthly temperatures for Alert on the north coast of Ellesmere Island in stark contrast with the South Pole station in Antarctica. Most of the limited precipitation falls as rain in summer. Annual precipitation totals are between 150 and 250 mm in high latitudes, but greater in alpine environments. Winter snowfall amounts are low and snow depths range from about 600 to 1600 mm. The region is wind-swept and blowing snow is common in winter. Snowmelt is rapid in June as a result of the long days, and the ground thaws to depths of 30–100 cm above the permafrost table. In consequence, the ground is

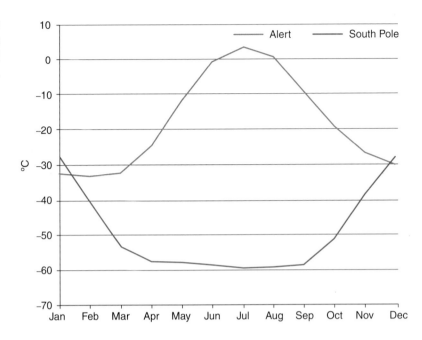

Figure 9.10
Mean monthly temperatures at Alert (82.5° N, 62.5° W) and South Pole Station.

frequently waterlogged. Permafrost is continuous beneath most of the Arctic tundra and has typical depths ranging over 100–600 m in Arctic Canada and Siberia.

(m) Ice plateaus

The two major *ice plateaus* are the Greenland and Antarctic ice sheets. Greenland is basically an ice dome rising to 3200 m at latitude 72° N. It extends 2400 km north–south and is over 1000 km wide at its widest point. Numerous small ice caps and glaciers surround it. Antarctica is 98 percent ice and covers 14 million km^2. It is made up of the larger East Antarctic ice sheet that rises to 4500 m and the smaller West Antarctic ice sheet flanked by the floating Ross ice shelf in the Ross Sea and the Filchner–Ronne ice shelf in the Weddell Sea. The ice shelves are composed of land ice that flows out onto the ocean; they are typically several hundred meters thick.

Greenland is affected by storm systems that enter Baffin Bay from the south, or move across it from the west. They also redevelop off the southeast coast, bringing significant precipitation in southeasterly onshore flow. The northern part of the ice sheet, however, is very dry. Mean annual accumulation on the ice sheet is ~340 mm water equivalent, with over 2000 mm in the southeast. The higher part of the ice sheet is dry snow, but near the margins there is substantial melt in summer. Ponds and small lakes form on the ice surface, some of which drain rapidly through moulins (shafts in the ice). At Summit Station

(73° N, 39° W, 3207 m) the winter mean temperature is −41 °C and the summer mean is −15 °C. Temperatures at coastal stations average 4–6 °C in summer.

The climate of Antarctica is the most severe on Earth. The temperature at South Pole Station, located at 2835 m altitude, varies between −29 °C in December–January and −59 °C in July–August (Figure 9.10), with extremes of −13 °C and −83 °C. Notice that, unlike Alert, the winter temperatures remain low for six months in a pattern referred to as a "coreless" winter – there is no core of cold. It is caused by the consistent negative radiation balance in the absence of solar radiation.

Vostok station (78° S, 107° E, 3488 m) on the East Antarctic plateau has recorded a world record low temperature of −89 °C. There are strong temperature inversions in the lowest 1000–1500 m and temperatures may be 25 °C or more higher at the top of the inversion than at the surface. In winter (April–October), Antarctica is encircled by extensive sea ice, covering ∼20 million km^2, that augments the low temperatures of the continent.

Storm systems from the Southern Ocean spiral in toward the continent, bringing cloud cover and gale-force winds that cause blowing snow, but little actual snowfall. The East Antarctic plateau is an ice desert receiving <50 mm of precipitation annually. In some parts of the ice sheet there are small areas of blue ice where no snow accumulates due to sublimation and katabatic winds. Wind speeds are often extreme at the coasts as a result of air drainage that accelerates downslope. Mawson station (69° S, 78° E) has recorded a peak gust of 69 m s^{-1}.

SUMMARY

Deserts occupy 25 percent of the continents. Annual rainfall is <250 mm. They exist due to subsidence and dry air in the subtropical anticyclones; interior basins remote from moisture sources; cold coastal currents; and cold polar deserts. The first type is exemplified by the Sahara and Arabia; the second by the Taklamakan and Gobi; and the third by Namibia and the Atacama. Cold polar deserts occupy the Arctic islands.

Monsoon climates with a seasonal wind reversal occupy 16 percent of the globe; they extend from East Asia/Indonesia westward to West Africa, with separate areas in northwest Mexico to the southwest United States and central South America. The wind reversal over India is strongly influenced by the Tibetan Plateau. In winter the subtropical westerly jet stream is locked in place by the Himalaya and Tibetan Plateau. In May–June the jet shifts north of Tibet and heating of the plateau sets up a tropical easterly jet stream over India. Monsoon rains are brought by depressions in the tropical easterlies from the Bay of Bengal and in the Western Ghats by low-level southwesterlies. Nearly 80 percent of Indian rainfall comes from the summer monsoon, which has active and break periods. The monsoon over China derives its moisture from Indonesia and the South China Sea. It reaches northwestern China by late July.

Over West Africa, monsoon air advances northward from the coast from February to May, then in mid-June rain occurs widely between 9° and 13° N. Most is associated with westward-moving waves, steered by two easterly jet streams, and north–south disturbance lines.

The North American monsoon is determined by the heating of the Colorado Plateau in June–July. Rains begin in June in Mexico and around 1 July in southern Arizona–New Mexico. Moisture surges from the Gulf of California affect low elevations. In South America, Amazonia is considered monsoonal, with its wet season in austral summer. The rains extend south and eastward as the season progresses. The principal moisture source is the South Atlantic. Over northern Australia, low-level westerly winds last from late December to mid-March. Rains come from monsoon depressions in the overlying tropical easterlies.

The Tibetan Plateau averages 4500 m elevation with several mountain ranges with glaciers. The northern half is underlain by permafrost. Precipitation decreases from 2000 mm in the southeast to 50 mm in the northwest. The southern margins are affected by the Asian summer monsoon. The much smaller Altiplano in South America is dry steppe. Rain falls in austral summer with moisture from the Amazonian lowlands.

The Amazon rainforest climate has a high degree of seasonal constancy in temperature and humidity. There is a continental heat low from October to April with substantial rainfall. About 25 percent of moisture is recycled over the basin; the rest is from the tropical Atlantic. In the east, instability lines develop between land breezes and the trade winds, and progress inland and bring rainfall. In austral winter, southerly cold surges may cause frosts in southern Brazil.

The equatorial Congo basin has rain seasons associated with the ITCZ and associated heat lows or waves in the easterlies. In January, the ITCZ runs north–south from 15° S to 5° N, between South Atlantic westerlies and Indian Ocean easterlies; along 15° S is an Inter-Ocean Convergence Zone, which in July extends from 15° S to the Equator. In March–April the rainy season is in the north. The rainfall maximum shifts south with the ITCZ to the central Congo in September–October and from November–February is in the southern Congo.

Tropical steppe in the Sahel has summer rains from westward-moving instability lines. It experiences periodic major droughts, with the late twentieth century being notable.

The southeast United States and southeast China have a humid subtropical climate. Winters are colder in China due to Siberian cold outbreaks. The southeast United States is dominated by tropical maritime air from the Gulf of Mexico. Inland in spring and summer thunderstorms and tornadoes are common. Southeast China has summer rainfall from the East Asia–West Pacific monsoon.

Mediterranean climates occur at about 35° latitude on west coasts, related to seasonal north–south shifts of the subtropical anticyclones. The classic wet, mild winter and hot, dry summer type around the Mediterranean Sea is limited to the central and eastern portion.

Temperate lowlands are both maritime and continental in the westerly wind belt; continentality increases, and precipitation decreases, eastward. All four seasons are well marked and there is considerable day-to-day weather variability.

Maritime west coasts are in the path of year-round westerly winds and traveling low-pressure systems giving cloudy mild weather. Exposed coasts and uplands tend to be very windy and coastal mountains may have high winter snowfalls.

Mid-latitude steppe and prairie is cold in winter and hot in summer. Most rain falls in summer convective events.

The boreal forest is the largest terrestrial biome. The climate is subarctic, with severe winters and short summers. Northeast Siberia and northwest Canada have persistent winter high pressure. There are substantial snow packs and much of the area has permafrost.

Tundra is north of the Arctic tree line and above the alpine tree line. There are long, cold winters and short, mild summers. There is low winter snowfall and most precipitation is summer rain. Permafrost underlies most of the region.

The Greenland and Antarctic ice sheets have very low temperatures and little snowfall, except at the coasts. There is winter darkness and summer daylight.

QUESTIONS

1 Summarize the major causes of deserts.
2 Compare the monsoon climates of the Indian subcontinent and West Africa.
3 Compare the equatorial climates of the Amazon and Congo basins.
4 What factors give rise to the extreme annual rainfall totals in Assam, Mt. Waileale (Hawaii), and the western slope of the Southern Alps, New Zealand?
5 Compare and contrast the climates of Greenland and Antarctica.
6 Explain the severe winter conditions in northeast Siberia.
7 Explain the variety of rainfall regimes around the Mediterranean Sea.
8 Contrast the climatic regimes in the southeastern United States and southeast China.
9 Compare the climatic conditions at the western and eastern ends of the Walker circulations.
10 Plot a hythergraph for your location or nearby weather station and compare it with the one for the closest climatic type in Chapter 9.

10 Past climates

Snow and ice-covered mountain in Antarctica.

ALTHOUGH GEOLOGISTS recognized the occurrence of past ice ages in the 1840s, the study of climate change only began in earnest in the 1950–1960s. Since the 1980s there have been major advances in our understanding of the nature of, and reasons for, climatic changes. These advances have been made possible through the development of new observational tools and hypotheses, and the use of global climate models to simulate past climatic conditions.

Knowledge of the climate of the past is of vital importance for a full understanding of present and potential future world climates. Analysis of past changes can help to reveal the underlying causes of global climate patterns. In this chapter we will summarize our knowledge of geologic time, the Cenozoic era, the Quaternary period, and the Anthropocene, when human influences on the environment became significant.

It is worth pointing out that *climate variability* refers to the variation of global or regional climate about its mean state. The time scale may be decades or millennia and the variation may be cyclical or irregular. *Climate change* refers to changes in the long-term statistics of climatic elements (the mean state, standard deviation, or frequency distribution). The time scale is at least several decades, but may be thousands of years or more. Climatic changes may be due to changes in external forcings (solar output, changes in the Earth's orbit, volcanic eruptions), internal variability in the atmosphere–ocean–cryosphere system), or anthropogenic (human-induced) forcing. Changes may involve an abrupt shift in the mean state, an increasing or decreasing trend over time, or a change in the degree of variation about the mean without a change in the mean value.

10.1 Geologic time

The Earth was formed about 4.5 billion years ago. It is thought that water was supplied by numerous cometary impacts, and by 4.2 billion years ago there were oceans. The young sun was about 25 percent fainter between 3.8 and 2.5 billion years ago than it is now (its luminosity increases about 6 percent per billion years) and it is probable that high levels of greenhouse gases (GHGs; carbon dioxide and methane) in the atmosphere prevented the water from freezing. The oldest rocks, found in Canada, date from 4.0 billion years ago. Figure 10.1 summarizes the *geological time scale*; note the three different time scales. Life at the molecular level emerged around 3.7 billion years ago and multicellular forms around 2.5 billion years ago, after which oxygen was formed in the atmosphere and the ozone layer developed. There were several severe cooling events in the Proterozoic (early life) era that are thought to have led to glaciations. The first occurred around 2.3 billion years ago, perhaps because oxygen in the atmosphere reduced the methane content, lessening the greenhouse effect. Two other glaciations that led to a state called *Snowball Earth*, with global land and sea ice cover, are dated to 710 and 640 million years ago (Ma), perhaps associated with the formation of the first supercontinent of Rodinia. Animal life appeared and evolved rapidly after the second of these ice ages.

Two supercontinents broke apart around 550 Ma and the geophysical mechanism of plate tectonics led the four new continents of Laurasia, Baltica, Siberia, and Gondwana to move around in a process known as *continental drift*. This was first suggested by Alfred Wegener in 1912, but not accepted by geologists and geophysicists because they had not identified a mechanism. Finally, in the 1960s the theory of *plate tectonics* was developed. Plate tectonics refers to the movement of seven or eight major plates in the oceanic and continental lithosphere over the Earth's surface at a rate of about 10 cm per year. The movement of one plate beneath another results in earthquakes and volcanic activity, as in the "ring of fire" that surrounds the Pacific Ocean. It is accompanied by sea-floor spreading

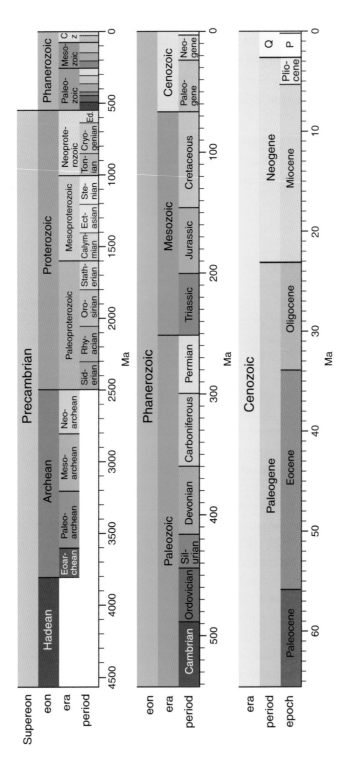

Figure 10.1 The geological time scale. The time scale for the Precambrian (top) is 4000 million years; for the Phanerozoic (middle) is 550 million years; and for the Cenozoic (bottom) is 66 million years in the lower right Q = Quaternary Period; P = Pleistocene epoch (source: Wikipedia, https://en.wikipedia.org/wiki/Geologic_time_scale).

with the formation of new crust, as along the Mid-Atlantic Ridge. The collision of plates gave rise to mountain building during many intervals of geologic time that would have modified the global atmospheric circulation and the distribution of climates. Global conditions in the early Paleozoic era ($\sim$ 550–400 Ma) were generally warm, except for a short ice age in the Ordovician around 470 Ma, when the continent of Gondwana was situated over the South Pole. There was prolonged glaciation in the Permo-Carboniferous period ($\sim$ 350–260 Ma) that ended when the supercontinent of Pangea formed. During the mid-Cretaceous period (around 100 Ma) high concentrations of CO_2 (up to 1600 ppm) led to a very warm and ice-free Earth and the age of the dinosaurs.

It is apparent that global climate changes during early Earth history were mainly determined by changes in solar luminosity, atmospheric composition and continental configuration.

10.2 The Cenozoic

We will now consider in more detail how world climate evolved during the Tertiary period (the start of the *Cenozoic Era*; Cenozoic is Greek for "new life") that began about 66 Ma (see the lower panel of Figure 10.1), after the age of dinosaurs when mammals evolved. Information about climatic conditions during this remote past is derived from sedimentary rocks that contain plant and animal fossils, and especially limestones that yield data on past CO_2 levels in the atmosphere. Conditions remained warm from 65 to 50 Ma with no ice sheets, perhaps due to high CO_2 concentrations. Between 55.5 and 52 Ma there were several abrupt and extreme global warming events. These have been linked to intervals with high eccentricity in the Earth's orbit and high obliquity (axial tilt) that triggered massive releases of soil organic carbon that had been locked up in permafrost in the Arctic and Antarctic. These pulses of carbon dioxide gave rise to the warming episodes.

Temperatures, especially in the deep ocean, then began to decline steadily with decreasing CO_2 levels. It has been suggested that this decrease was caused by prolonged chemical weathering of rock surfaces in the continent of Antarctica. About 33 Ma, the climate of Antarctica changed from mild and temperate to glacial conditions, with the build-up of extensive land ice on East Antarctica. It is uncertain as to whether this was caused by CO_2 levels dropping below a critical threshold, or whether it was perhaps a result of the opening of the Drake Passage that allowed the formation of the Antarctic Circumpolar Current; this would have excluded warm waters from the seas around the continent.

By about 17 Ma the continents had drifted essentially into their present locations, establishing the modern patterns of circulation in the atmosphere and oceans.

In the Middle Miocene, about 18–16 Ma, there was a warm period known as the Miocene Climatic Optimum, probably caused by high CO_2 concentrations in the atmosphere. Mean annual temperatures in Central Europe reached 22 °C, compared with ~ 10 °C today. Then the Earth cooled rapidly and by ~ 15 Ma there was already a large Antarctic ice sheet, but the Isthmus of Panama had not yet closed and the different ocean circulation inhibited the growth of land ice in high northern latitudes. However, since about 14 Ma there appears to have been perennial Arctic sea ice. Glaciation in the Arctic (Greenland) appears to have begun around 3 Ma, when several global cooling steps occurred, related to atmospheric carbon dioxide levels, and the Panama seaway finally closing around 2.9 Ma. This set up the present pattern of ocean circulation in the North Atlantic. The mechanisms involved in the inception of Arctic glaciation are still not well understood.

10.3 The Quaternary

The *Quaternary Period* (the Pleistocene and Holocene epochs) is considered to have begun about 2.6 Ma. Its importance lies in the widespread and repeated appearance of major *glacial cycles*. Evidence of these is provided by (1) sediment cores from the ocean floor that contain marine fossils, indicating ocean temperatures and oxygen isotope records of past ice volume; (2) long pollen records of past vegetation; (3) the annual stratigraphy of ice core records from Greenland and Antarctica that indicate snow accumulation, atmospheric temperatures from oxygen isotopes, and atmospheric chemistry; and (4) glacial deposits demarcating the former extent of glaciers. Details of these sources of evidence are presented in R. S. Bradley's *Paleoclimatology: Reconstructing Climates of the Quaternary*, but here we give a brief overview of the major types of proxy records.

Proxy records are chemical, geological, and biological indictors of climatic conditions that provide long time series of particular climatic elements before instrumental observations became available. Examples of proxies include the chemistry of deep ice cores, tree rings, borehole temperatures, coral growth rings, paleoglacial features, and lake and ocean sediments containing pollen and microfossils.

One of the most powerful chemical proxies is the oxygen isotope record in ice cores and marine microfossils. Molecular oxygen is mainly made up of the light isotope ^{16}O, but a small fraction is the heavier ^{18}O isotope. When water evaporates from the ocean, it preferentially contains the lighter isotope so the seawater is enriched in ^{18}O by 2–3 ppm during glacial cycles when water is transferred from the oceans to ice sheets. These variations are registered by planktonic and benthic foraminifera that are deposited when they die in marine sediments. Additionally, as air flows inland over Greenland or Antarctica, precipitation is progressively depleted of the heavier isotope, giving increasing deficits of ^{18}O (−30 to −50 ppm) that indicate the progressively lower condensation temperature in the air. The seawater $\delta^{18}O$ enrichment increases during glaciations, when there is less water in the ocean, giving a measure of the volume of land ice. Hence, the timing and extent of past glacial ages can be tracked. Marine sediments have a time resolution of one to several thousand years. Ice cores have annual layers and the $\delta^{18}O$ traces air temperature at the time of condensation of the cloud water. These ice cores span up to several hundred thousand years. Glacier ice also traps air bubbles and this enables the past atmosphere to be analyzed in terms of its carbon dioxide and methane contents. Other elements retained in ice include sea salts, heavy metals, and volcanic sulfates. The last of these is determined by measuring the electrical conductivity along the ice core.

Another chemical proxy method investigates speleothems, or cave deposits such as stalactites and stalagmites, of limestone, which can also be analyzed for their oxygen isotope content because limestone is calcium carbonate ($CaCO_3$) or magnesium carbonate ($MgCO_3$). The growth rings in these formations, when dated by uranium-thorium methods, give information about past temperature and moisture conditions. Uranium-thorium dating is based on the detection by mass spectrometry of both the parent (^{234}U) and daughter (^{230}Th) products of decay. Dates back to 300 000 years can be obtained.

Terrestrial boreholes contain a temperature profile in the ground that preserves the integrated effect of past changes at the surface. Mathematical analysis is used to reconstruct these past conditions. Temperatures about 100 years ago are preserved at 150 m depth and 1000 years ago at 500 m depth.

Annual growth rings in trees and corals can be used to estimate past temperature and/or moisture conditions. Corals retain more ^{18}O in colder and more saline waters. However, records only span 50–60 years. Tree rings span several thousand years. Ring widths are narrow or absent in dry conditions at the lower timberline and in cold summers at the upper timberline. Another indicator in tree rings is the maximum late-wood density, which is measured by X-ray densitometry (see Box 10B.1).

Pollen trapped in peat bogs and lake sediments provides information on the regional vegetation cover. Over the Holocene the temporal resolution is about 10–100 years. Pollen grains from trees, plants, and grasses can be identified, usually to genus level. Several hundred pollen grains are identified and counted on a slide taken from each successive layer. The chronology is mainly determined from radiocarbon (^{14}C) dating of charcoal in the profile that can extend back to about 60 000 years. Some pollen records extend back for 500 000 years.

Other biological techniques include the typing of midge and beetle assemblages on land and diatoms in lakes and the ocean that are often temperature and

Box 10B.1 Dendrochronology and dendroclimatology

Dendrochronology is the study and dating of annual rings in trees. Tree-ring analysis assumes that annual growth rings reflect significant events during the tree's life. Growth rings form in the xylem wood of the trees. Early in the season, xylem cells are smaller and darker. An abrupt change from light to dark rings delineates the annual increments of growth. Studying the size and variations in the rings provides information about the varying environmental conditions. Tree-ring growth is influenced by a range of environmental factors, the most important being climate.

Under conditions of stress, growth is retarded and tree rings narrow; under more favorable conditions, wider annual rings are produced. As a result, climatic variations over shorter but precisely dated time scales can often be inferred, an area of study known as *dendroclimatology*. The importance of this precise dating technique is that inferences can be made about past seasonal variations.

A. E. Douglas and his colleagues at the University of Arizona pioneered the study of tree-ring analysis. Initial studies attempted to relate seasonal growth of trees to sunspot cycles. Analysis of ancient living trees, such as the *Pinus aristata*, found to be 4000–5000 years old, permitted reconstruction of climates of the American southwest during various periods. This analysis method is most valuable in determining conditions that existed during the last few millennia and is used extensively in archaeology.

moisture dependent; and the analysis of plant macrofossils (leaves and seeds) from pack rat middens in the arid southwest United States.

Geomorphological methods include the mapping of former lake shorelines that indicate past water budgets and of glacial moraines and periglacial features that indicate past ice extent. There is abundant evidence in many areas for variations in sea level during the Quaternary. This includes former coastal landforms now standing above present sea level, such as former beaches, deltas, spits, and coral reefs. Evidence for sea-level change can also be found offshore in the form of submerged landforms. Mapping of such features enables the positions of former coastline to be established and the vertical range of sea-level variations to be estimated.

It has long been recognized that landforms of glacial deposition and erosion are important paleoenvironmental indicators. When carefully mapped, they reveal a great deal about the extent, thickness, and behavior of former ice masses, the direction of ice movement at both local and regional scales, and the nature and pattern of glacier retreat. In some cases the information may be used to reconstruct the configuration of the former ice sheet or glacier surfaces, and to enable former ice volumes to be estimated. Comparisons with present-day glaciers, where a close relationship between glacier behavior and climatic parameters has been established, enables inferences to be made about former climatic regimes. This evidence is a key element in the development of computer models of the geometry of former ice sheets and glaciers.

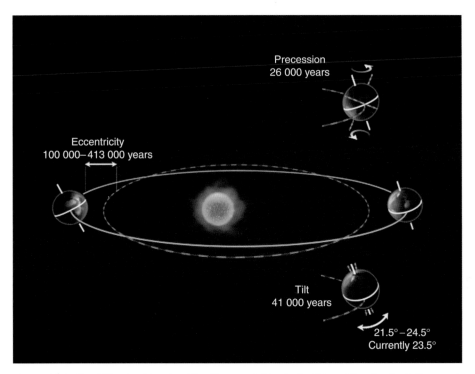

Figure 10.2 The Milanković cycles. The three variables in Earth's orbit, axis of rotation, and seasonal timing, with periods of variation marked (source: COMET® at the University Corporation for Atmospheric Research (UCAR) pursuant to a Cooperative Agreements with the National Oceanic and Atmospheric Administration, US Department of Commerce. ©1997–2009 University Corporation for Atmospheric Research).

The history of the Quaternary is intimately connected with variations in the *Earth's orbit* about the sun, the axial tilt of the Earth, and the position of the Earth at the spring equinox. The path of the Earth around the sun is slightly elliptical, and the degree of this orbital eccentricity varies over about a 100 000-year cycle (Figure 10.2). The axial tilt (formally known as the obliquity of the ecliptic) varies over a 41 000-year cycle. Finally, the position of the Earth at the spring equinox varies; the precession of the spring equinox, which alters the timing of the minimum (perihelion) and maximum (aphelion) sun–Earth distance approximately every 21 000 years due to the combined effect of precession and orbital eccentricity. Currently, perihelion occurs on 3 January, when the Earth is 5 million kilometers closer to the sun than on 4 July (aphelion). This distance effect causes the Earth to receive 7 percent more solar energy at perihelion than at aphelion. Around 10 000 years ago, when the Earth was closest to the sun in July, summers in the Northern Hemisphere were warmer than now (Figure 10.3). In 10 500 years' time perihelion will again be in July, making northern summers warmer, especially in the Subtropics (Figure 10.3).

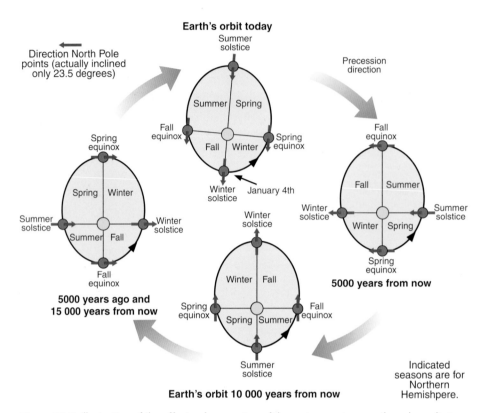

Figure 10.3 Illustration of the effects of precession of the spring equinox on the solar radiation receipts in the Northern Hemisphere at different stages of the cycle: 5000 years ago, today, 5000 and 10 000 years from now (source: Wikipedia, http://en.wikipedia.org/wiki/ Milankovitch_cycles).

Until about 0.8 Ma, cold intervals had a 41 000-year (41 ka) periodicity coincident with the period of variation of the Earth's axial tilt (or obliquity), which varies between 22.1° and 24.5°; it is currently 23.5°. The obliquity effect is equal in the two hemispheres and its intensity increases poleward. There is increased seasonality when the tilt is larger. For unknown reasons, after 0.8 Ma the periodicity switched from 41 ka to ~100 ka, coincident with the eccentricity of the Earth's orbit around the sun. However, the fluctuations in solar radiation due to this eccentricity are too small in themselves to trigger glacial onset without amplification by ice-albedo and other feedbacks. The eccentricity of the Earth's orbit also shows a longer periodicity of about 400 000 years and this may have been involved in the switch at 0.8 Ma.

The axial tilt and precession of the equinox together give rise to seasonal variations in incoming solar radiation of about 10 percent, while the effect of eccentricity is only about 0.2 percent. The three orbital effects were first described

Box 10B.2 Milutin Milanković

Milanković was born in Croatia (then part of Austro-Hungary) in 1879. After an early career as a civil engineer in Vienna he was appointed Professor of Applied Mathematics in Belgrade in 1909.

Interned during World War I in Budapest, he was able to work on the mathematical theory of climate and published a book on this topic in 1920. He continued to work on the relationship between solar radiation and the Earth's climate, and in 1941 collected his works together in a book published in Serbian that was translated into English in 1969 under the title *Canon of Insolation of the Ice-Age Problem*. He died in 1958, almost two decades before his theories on the role of astronomical variations in global climate change became widely accepted by meteorologists.

quantitatively by the Serbian astronomer Milutin Milanković in 1920 (see Box 10B.2), but the concept of astronomical variations being a major forcing for ice ages was only given serious attention in the 1970s in the work of A. D. Vernekar, Andre Berger, J. Hays, J. Imbrie, and N. Shackleton. The last three authors identified the orbital variations as being "the pacemaker of the Ice Ages."

EPICA's (the European Project for Ice Coring in Antarctica) ice core collected from Dome C in Antarctica has been analyzed back to 0.8 Ma, recording eight glacial cycles (Figure 10.4). Until 430 ka the interglacials occupied a larger fraction of each cycle, but were less warm than the subsequent ones (see Box 10A.1). During each glacial cycle, glacial conditions last about 90 percent of the time, and interglacials only 10 percent. The interglacials saw sea levels higher than at present; during Marine Isotope Stage 11 (427–365 ka), sea levels rose ~6–13 m above present and both the Greenland and West Antarctic ice sheets collapsed, according to Raymo and Mitrovica. Lake sediment records from northeastern Siberia indicate that stage 11c was 4.5 °C warmer than in the Holocene. In the last Eemian Interglacial that peaked around 125 ka, new work puts the sea level at least 6.6 m above today, according to Kopp *et al.*, implying that the West Antarctic ice sheet disappeared. Forests expanded northward in Europe and North America and temperatures were probably 1–2 °C above present.

Beginning around 115 ka the Earth began to cool; there was extensive glaciation in both hemispheres by 75 ka. Following some amelioration, more severe glacial climatic conditions returned after 30 ka. During the *Last Glacial Maximum* (LGM), which culminated about 21 ka, land and sea ice covered about 30 percent of the Earth. Ice sheets 3–4 km thick covered northern North America (the Laurentide, which extended from Canada into the northern United States; the Cordilleran over the western cordilleras; and the Innutian in the Queen Elizabeth Islands of Arctic Canada); and Fenno-Scandinavia, with ice also over northern and western Britain

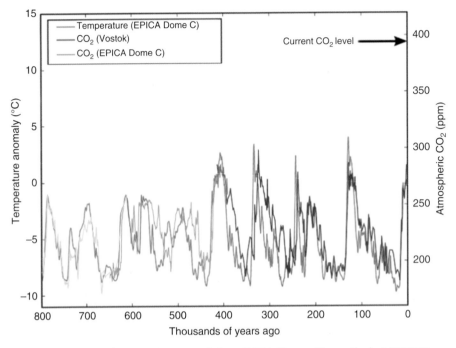

Figure 10.4 CO$_2$ and temperature records from EPICA (Dome C) over the last 800 000 years and Vostok for the last 430 000 years. The modern CO$_2$ level is shown for comparison (source: Wikipedia, http://en.wikipedia.org/wiki/File:Co2-temperature-plot.svg).

and the North Sea, and a marine-based ice sheet in the Barents Sea. However, in northern Asia there was only local ice cover due to the extreme dryness. Average temperatures fell by almost 6 °C globally, about 5 °C in the Tropics and 15 °C in mid-latitudes. The presence of the massive Laurentide ice sheet resulted in a split jet stream, with storm tracks passing both north and south of the ice sheet. Permafrost underlay Siberia and much of central Asia and Europe. The sea level dropped by about 130 m due to the volume of water locked up in the ice sheets, exposing vast continental shelves in the Arctic Ocean, Bering Sea, North Sea, and the seas between Australia and Indonesia. Snowlines were lowered by about 1000 m, on average, and mountain glaciers were much more extensive than today.

Global climate underwent large and abrupt fluctuations during the last glacial cycle. Ice cores from Greenland record 24 such Dansgaard–Oeschger oscillations since 105 ka, with a recurrence time of about 1500 years. They are characterized by a rapid, decadal-scale warming followed by a slow, centuries-long cooling. Their cause is still unclear, but they appear to reflect changes in the ocean circulation of the North Atlantic.

Box 10A.1 Marine Isotope Stages and interglacials

Marine Isotope Stages (MIS) are a numbering system for glacial and interglacial intervals developed by Cesare Emiliani in the 1950s and Nicholas Shackleton in 1967. Odd numbers are interglacials, beginning with MIS-1 for the present Holocene period; even numbers are glacials, beginning with MIS-2 for the last glaciation that started around 24 ka. They are determined from oxygen isotope records (^{18}O) in marine foraminifera (plankton species) that are largely made up of calcium carbonate ($CaCO_3$). When they die, the forams sink to the seabed and are incorporated in the marine sediments. High ^{18}O in marine organisms represents cold intervals and low ^{18}O warm intervals.

Yin and Berger model the contribution of GHGs and incoming solar radiation (insolation) to the climate of past interglacials. They find that GHGs play a dominant role in the variations of the annual mean temperature of both the globe and southern high latitudes, whereas insolation plays a dominant role in the variations of precipitation and of northern high latitude temperature and sea ice. The relative importance of GHGs and insolation for the intensity of warmth varies from one interglacial to another. For the warmest (MIS-9, 334–302 ka, and MIS-5, 130–72 ka) and coolest (MIS-17, 712–660 ka and MIS-13, 528–475 ka) interglacials, GHGs and insolation reinforce each other. MIS-11, 427–365 ka (MIS-15, 621–569 ka), is a warm (cool) interglacial due to its high (low) GHG concentration, despite its insolation contributing to a cooling (warming). MIS-7 (244–191 ka), although having high GHG concentrations, cannot be classified as a warm interglacial due to its large insolation-induced cooling. MIS-19 (787–761 ka) appears to be the best analogue, in terms of these two forcings, for the Holocene MIS-1.

Recent work suggests that the post-glacial warming from 22 ka was initiated by increasing axial tilt and the precessional shift of the June solstice toward perihelion, which warmed high northern latitudes and began to melt the ice sheets. This was accelerated by rising CO_2 levels (supplied by the Southern Ocean) that have been shown to precede global temperature increases by a few centuries. The rise in Southern Hemisphere temperatures is steady, while those in the Northern Hemisphere dip around 17 ka and 12.5 ka as a result of a weakening of the Atlantic meridional overturning circulation. The general warming trend after the LGM was interrupted by the abrupt *Younger Dryas* cold interval around 12 800–11 500 calendar years ago when glacial-like conditions briefly returned, in association with changes in ocean surface temperatures. It is thought that these were the result of a significant reduction or shutdown of the North Atlantic's meridional overturning thermohaline circulation in response to a sudden massive influx of freshwater from glacial Lake Agassiz (near the modern Great Lakes) – which was dammed up by the Laurentide ice sheet – and the deglaciation of North America that resulted in massive iceberg calving.

Box 10B.3 Dating the past

Before Present (BP) refers to radiocarbon dates determined by the decay of carbon-14 in organic matter, including charcoal. The reference date used is 1950. Carbon-14 has a half-life of 5570 years, meaning that half of the initial amount has decayed in that time interval. Typically there will be a series of dates for a given core to bracket different climatic phases. Radiocarbon dating can be applied back to about 50 000–60 000 years.

Dating of older materials is usually performed using uranium and thorium isotopes (^{238}U, ^{234}U, ^{232}Th, and ^{230}Th) in corals, for example. ^{234}U has a half-life of 245 ka and ^{230}Th of 75 ka.

Freshwater is less dense than seawater and so forms a stable upper layer in the ocean, preventing vertical overturning. A contributory effect of the freshwater influx would have been the increase in sea ice cover, since that water freezes at 0 °C instead of the −1.8 °C for seawater.

During the last glacial cycle in Africa there was a cold, arid phase about 70 ka, followed by a slight climatic amelioration and then a second aridity maximum around 22–13 ka. These arid phases coincided with extensive glaciation in more northern latitudes. Conditions then quickly became warmer and moister, leading up to the Holocene "optimum" of greater rainforest extent, and savanna vegetation covering the Sahara, with high lake levels from 9500–6500 Before Present (BP) (see Box 10B.3). This allowed the expansion of population into areas of North Africa that are now desert. Conditions then turned much more arid around 5500 BP in the Sahara and the Thar desert of northwest India, becoming similar to the present.

The start of the *Holocene* (post-glacial time) is now dated to 11 700 BP, although the Fenno-Scandinavian ice sheet disappeared only 9000 years ago and the Laurentide about 6500 years ago. The final remnant of the Laurentide ice sheet appears to be the present-day Barnes ice cap on Baffin Island. Pollen records from lakes and peat bogs that allow reconstruction of the vegetation, beetle assemblages that indicate microclimate, and pack rat middens in North American desert environments that contain plant remains, provide evidence of Holocene paleoclimate on centennial time scales. Annual time scales can be resolved from annual growth rings in trees and corals, as well as annual layers in shallow ice cores from the major ice sheets and tropical mountain ice caps.

Conditions during most of the Holocene were much less variable than during the glacial periods. Thermal maximum conditions (the Hypsithermal) are dated to 8000–5500 BP over much of the globe, but around 9000 BP in high latitudes away from the remnant Laurentide ice sheet. The thermal maximum is attributed to increased solar radiation receipts resulting from the occurrence of perihelion (Earth closest to the sun) in July 10 000 years ago, rather than in January as at present (Figure 10.3). This made boreal summers warmer than now and up to 5 °C warmer in northern and southern high latitudes. It also led to enhanced monsoonal circulations in South Asia and West Africa as a result of the warmer landmasses. Between 10–8 ka and 7.3–6.4 ka, mega-lake Chad in West Africa covered an area of 350 000 km^2 between 11° and 18° N, compared with 1500 km^2 in 2000.

This warm interval was followed by a cooling trend with neoglacial episodes of glacier re-advance, notably during the *Little Ice Age* (LIA), dated to AD 1550–1850 in Europe and around the North Atlantic, or AD 1400–1700 for the Northern Hemisphere as a whole according to Mann *et al*. Temperatures were about 0.5–1 °C below those at present, but fluctuated widely on decadal time scales.

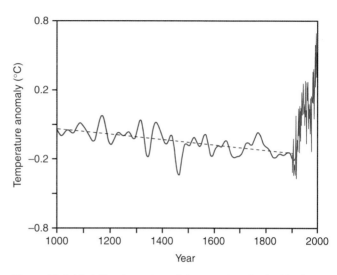

Figure 10.5 Variation in surface air temperature in the Northern Hemisphere over the past millennium. The reconstructed 40-year smoothed values are plotted for AD 1000–1880 (blue), together with the linear trend (dashed) for AD 1000–1850 and observed temperatures (purple) for AD 1902–1998. The reconstruction is based on estimates from ice cores, tree rings, and historical records and has two standard error limits of about ±0.5 °C during AD 1000–1600. The values are plotted as anomalies relative to AD 1961–1990 (source: adapted from Mann *et al*. 1999; Courtesy of M. E. Mann, Pennsylvania State University).

The LIA succeeded a brief mild interval known as the *Medieval Warm Period* (or Medieval Climate Anomaly) around AD 950–1250. Temperatures were about 0.5–1.4 °C above those of 1961–1990 for the northwestern and central North Atlantic, the southern United States and central China. During this interval the Vikings sailed the North Atlantic and built settlements in southwest Greenland that later succumbed to the deteriorating climate. The general pattern of hemispheric temperature over the last millennium is shown in Figure 10.5. The trend resembles a hockey stick, with an abrupt rise at the end. Over the last 1000 years there have been persistent dry periods lasting up to several decades over North America, West Africa, and East Asia. These historical mega-droughts are linked to tropical sea-surface temperature (SST) variations, with La Niña-like SST anomalies in the tropical Pacific often leading to widespread drought in North America according to Diaz *et al*., and El-Niño-like SST warming in the Pacific causing drought in eastern China.

Other likely causes of climatic variations over the last millennium are fluctuations in solar output and volcanic activity. Solar output (across all wavelengths) varies about 0.1 percent over the < 11-year solar cycle. In 1976, Jack Eddy identified the near-absence of sunspots from 1645 to 1715 (the so-called Maunder Minimum) as a potential cause of the LIA, a hypothesis that was later largely accepted. Solar output is increased during sunspot episodes due to the increased emission from the brighter solar faculae, which outweighs the decrease from the darker (and cooler) sunspots. During the Maunder Minimum, which coincided with the middle of the LIA, there was a prolonged absence of sunspots. However, it is not known how much total solar irradiance decreased during that time; estimates vary from < 0.3 percent to as little as 0.1 percent. There is debate as to how the solar cycle affects global temperatures. One theory involves "top-down" effects via stratospheric ozone. Another suggests "bottom-up" effects on evaporation over tropical oceans that are transmitted to the troposphere via latent heat in convection.

Explosive *volcanic eruptions* in the Tropics inject particles and sulfur dioxide into the stratosphere, and this reduces summer temperatures in Europe by $\sim 0.5\,°C$ in the first two summers after the event, based on the analysis of 15 eruptions over five centuries. Winter temperatures are increased by up to 2 °C. It has recently been proposed by Miller *et al.* that four volcanic eruptions in the late thirteenth century caused cooling that led to sea ice expansion in the Arctic and triggered the onset of the LIA in Iceland and Baffin Island. Further volcanism in the mid fifteenth century gave this cooling a boost. In Scandinavia and the Alps the expansion of valley glaciers led to direct loss of farmland, while the low temperatures often resulted in reduced crop production and harvest failures.

Late in the LIA the cataclysmic eruption of Mount Tambora, Indonesia, in April 1815 lowered global temperatures by as much as 3 °C. Most of the Northern Hemisphere experienced sharply lower temperatures during the succeeding summer months. In parts of Europe and eastern North America, 1816 was known as "the year without a summer." Similar effects were noted following the eruption of Krakatoa in Indonesia in 1883. The temperature decreases are attributed primarily to sulfate aerosols in the stratosphere that circled the globe, as well as to changes in cloud cover. Because the eruption was at the Equator (like Tambora), the ash and aerosols spread into both hemispheres.

10.4 The Anthropocene

The *Anthropocene* is a term that has recently been proposed by Zalasiewicz *et al.* for the period when human influences on global climate have become widely apparent. The start date of this interval is uncertain, but it coincides approximately with the beginning of the Industrial Revolution around AD 1800. Climatic observations in many parts of the world begin about this date and there are also older documentary records (weather diaries, records of floods and severe winters, wine production and harvest data). Much of this information was assembled by the late Hubert Lamb in his book *Climate: Past, Present and Future*, published in 1990. Lamb began work on climatic changes and its causes in the 1960s, along with other notable climatologists such as Gordon Manley, Hermann Flohn, and Murray Mitchell (see Box 10B.4). Manley assembled a record of central England temperatures (CET) from the mid seventeenth century and an analysis of the data spanning 335 years by Plaut *et al*. found oscillations of 7–8, 15, and 25 years. They suggested that the multidecadal oscillations were related to variability in the atmosphere and ocean of the North Atlantic. Recent work suggests that cold winters in the CET record may be linked to sunspot minima. This relationship operates on a regional scale for European winters.

In the nineteenth century concentrations of atmospheric carbon dioxide, methane, and pollutants began to rise. Svante Arrhenius, in 1896, and George Callendar, in 1938, were the first to point out the *"greenhouse effect"* on atmospheric temperatures due to the increasing absorption of infrared radiation by GHGs in the atmosphere (see Box 10A.2). The Intergovernmental Panel on Climate Change (IPCC) issued its Fourth Assessment Report in 2007; the scientists involved were awarded the Nobel Peace Prize in that year. The report concluded that there was overwhelming evidence that the global climate was warming, almost certainly as a result of human activity increasing the atmospheric load of GHGs. Mean global temperature rose by 0.7 °C in the twentieth century and by twice as much in the Arctic. From the 1980s, Arctic sea ice extent declined to a record low in September 2007 (nearly equaled in 2008, 2010, and 2011), and then surpassed that low value in September 2012 with an extent of only 3.4 million km^2, less than half of the 1979–2000 average value. Almost all mountain glaciers worldwide shrank substantially from their advanced positions at the end of the LIA (in the late nineteenth century), especially after 1980.

From AD 1800, *carbon dioxide concentrations* in the atmosphere increased by almost 40 percent from ~ 280 ppm to 390 ppm (at 2012) due to fossil-fuel burning and changes in land use, while methane concentrations rose 2.5 times from ~ 700 ppb to 1825 ppb due to agriculture (paddy rice and cattle), termites, and release from wetlands and landfill sites. *Methane* is a more potent GHG than carbon dioxide. It has a lifetime in the atmosphere of only 12 years, but compared with CO_2 (value 1), methane has a greenhouse warming potential value of 25 over a century. For comparative purposes, we may note that Antarctic ice cores (see Figure 10.3) show that, in response to fluctuations of CO_2 concentration from 180 to 280 ppm during glacial cycles, air temperatures fluctuated by ~ 8 °C!

Box 10A.2 The history of research on greenhouse gases and global warming

In the 1820s, Joseph Fourier in France recognized that atmospheric gases could trap energy from the sun and make the Earth warmer than it otherwise would be.

In 1859 John Tyndall in England identified water vapor and CO_2 as having the capacity to trap heat in the atmosphere and being most effective in offsetting nighttime cooling. The first scientific statement of the role of GHGs in the atmosphere was made by Svante Arrhenius in Sweden. In 1896 he calculated that reducing the CO_2 concentration in the atmosphere by half would lower the air temperature by 4–5 °C. The amplifying effect of ice-albedo feedback in high latitudes would then permit ice age conditions to develop. Conversely, increasing CO_2 would raise temperatures globally.

Arrhenius also recognized that since warmer air can hold more water vapor, there would be a positive feedback effect that would further raise air temperature since more water vapor would lead to more trapping of infrared radiation. Although water vapor accounts for about half of the natural greenhouse effect, it is CO_2 that is the controlling factor since it remains in the atmosphere for long intervals, whereas water vapor is constantly recycled. Arrhenius estimated that a doubling of the CO_2 concentration would raise global temperature by 5–6 °C, close to modern estimates. This idea was dismissed for 50 years, although in 1931 an American physicist, E. O. Hullbert, got results similar to Arrhenius in a little-noticed paper.

In Britain, G. S. Callendar in 1938 calculated that atmospheric CO_2 concentration had increased by 10 percent since the nineteenth century and that this could explain the observed warming since the 1890s. He also estimated that a future doubling could raise temperatures by 2 °C.

In 1956, using a better atmospheric model of infrared radiation absorption and computers, American physicist G. N. Plass calculated that doubling CO_2 would raise temperatures by 3–4 °C. In 1957 H. E. Suess and Roger Revelle drew attention to increasing levels of CO_2, as did Callendar in 1958. Beginning in 1957, C. D. Keeling began to measure CO_2 in the atmosphere every ten seconds on the top of the volcano Mauna Loa in Hawaii. The isolated location in the central Pacific provided an ideal benchmark of the global signature.

Keeling soon detected a steady increase superimposed on an annual cycle due to the seasonal change in plant biomass. The Mauna Loa record and those at numerous other locations around the world continue to be maintained.

In the 1970s and 1980s attention turned to other greenhouse gases (methane, nitrous oxide, ozone, and chlorofluorocarbons). In 1980, V. Ramanathan estimated that these trace gases could add 40 percent to the warming effect of CO_2. In 1985 a French–Soviet team led by Claude Lorius produced a 150 000-year ice core record from Vostok, Antarctica, that showed large fluctuations in past CO_2 levels (from 180 to 280 ppm) that correlated closely with temperature changes of the order of 8 °C (see Figure 10.4).

From 1990 through 2007, four assessments by the IPCC have refined the estimates of global warming by 2100 to 2.5 ± 1 °C, relative to 1980–1990, with a CO_2 concentration that will reach about 700 ppm. The IPCC Fifth Assessment Report (5AR) was released in September 2013.

Box 10B.4 Eminent climatologists

Gordon Manley was born in 1902 in Douglas, Isle of Man. In the 1920s he was a lecturer in geography at Birmingham University and then in 1931 at Durham, where he became Director of the Observatory. He installed a weather station at 847 m on Great Dun Fell, the first mountain station in England. He studied the easterly helm wind, identifying a standing wave and rotor. In 1952 he published *Climate and the British Scene*, an accessible work for non-academic readers.

From 1948 to 1964, Manley was Professor of Geography at Bedford College for Women in the University of London. Major contributions were his papers on CETs from the seventeenth century, published in 1953 and extended in 1974; the UK Meteorological Office regularly updates the time series. He died in 1980.

Murray Mitchell was born in 1928 in New York City. In the 1950s, from aircraft flights north of Alaska, he identified Arctic haze as being caused by air pollution from industrial areas in Eurasia and China. In 1965 he became a meteorologist in the Environmental Science Services Administration (later NOAA). He was a pioneer in research on climate change, particularly the effects of aerosols and GHGs and in the use of statistical methods. He died in 1990.

Hermann Flohn was born in Frankfurt, Germany in 1912. He published extensively on tropical climate, the Asian monsoon, and teleconnection patterns. In 1941 he wrote the first paper in German on human influence on climate. He was Head of the Meteorology Department at the University of Bonn in Germany. He wrote a number of books in both German and English, including one on the climate of Europe. He died in 1997.

It is clear that global climate is not stationary. However, non-stationarity can take several forms. There may be a gradual or abrupt shift in mean conditions, or there may be change in the degree of variability such that the distribution of extremes is altered. This can occur with or without a change in the mean value. An increased frequency of hot/cold or wet/dry days can lead to a warmer/colder or wetter/drier climate. A further possibility is that there are quasi-periodic oscillations without a change in the mean state over a long period.

The frequency of circulation patterns may change over time and this could reflect natural variability in the atmosphere or it could be a response to global warming. Using pressure data from 20 stations over Europe (30–70° N, 30° W–40° E), a zonal pattern and a blocking/cyclonic pattern have been identified by Slonosky *et al.* During 1822–1870, the circulation was more meridional with blocking/cyclonic patterns, whereas for 1947–1995 it was more zonal.

The *radiative forcing* due to anthropogenic GHGs since about 1800 is 2.3 W m^{-2}. This is made up of contributions from carbon dioxide, methane, nitrous oxide, halocarbons, and tropospheric ozone. Aerosols (sulfate, organic carbon, nitrate, and dust) give rise to cooling through direct radiative effects and indirect effects on cloud albedo as a result of cloud droplet condensation, with a total radiative forcing of -1.2 W m^{-2}. However, the effects of black carbon deposited on snow lead to a small positive contribution. Changes in land use give rise to a cooling effect as a result of increases in albedo. The total net radiative forcing from all sources is estimated currently to be 1.6 W m^{-2}.

Natural climate variability is attributable to solar variations, volcanic eruptions, and randomness in the atmosphere–ocean system. Solar variability over the $\sim$11-year sunspot cycle is at most 1 W m^{-2}, which gives only 0.25 W m^{-2} when averaged over the Earth. Volcanic eruptions releasing sulfate aerosols lower hemispheric or global temperatures, depending where the eruption occurs, by up to $\sim$1 °C in summer for 1–2 years after the event, but there were few major eruptions in the twentieth century.

Temperatures in the twentieth century rose to an initial peak in the 1930s and 1940s, then declined or leveled off, and have risen continuously since the mid 1980s. The warmest years on record occurred in the late 1990s and 2000s. The early to mid century warming, which was most marked in the northern North Atlantic, is potentially attributable to natural variability. The decline in Northern Hemisphere temperatures in the 1960s–1970s is attributed to higher sulfate aerosol levels in the atmosphere due to industrial activity. Over the period 1950s–1980s there was a *global dimming*, with a reduction in direct solar radiation by 4–6 percent attributed to this effect. Surface solar radiation decreased by 3–7 W m^{-2} per decade in major land areas. A *global brightening* ($+2$–8 W m^{-2} per decade) succeeded this during the 1980s to 2000, except in India, and the trend has continued in the United States ($+8$ W m^{-2} per decade) and Europe ($+3$ W m^{-2} per decade) as a result of controls on air pollution. In India the negative trend continues (-10 W m^{-2} per decade) and in China/Mongolia the trend turned negative after 2000. The brightening has contributed to the post-1980 global warming according to M. Wild.

Since 1980, global temperatures have risen by 0.5 °C, with the three warmest years on record (since 1850) being 2010, 2005, and 1998. The Northern Hemisphere land surface has warmed by about 0.1 °C for the years since 2001. The land surface in the Northern Hemisphere/Southern Hemisphere warmed by 1.12 °C/0.84 °C over the period 1901–2010, according to Jones *et al.* In the Arctic, temperatures have increased at almost twice the global average rate in the past 100 years as a result of *polar amplification*, involving the effects of ice-albedo positive feedback. This operates through a decrease in snow and ice cover, exposing bare ground or ocean with much lower albedos. The consequent increase in absorbed solar radiation raises the temperature, leading to further decreases in snow and ice cover. This effect, together with the advection of warm water and warm air from middle latitudes, gives rise to the increased Arctic warming as discussed by Serreze and Barry. Significant warming has also been observed in the Antarctic Peninsula, where air temperatures have risen by about 3 °C since the 1950s. This led to the disintegration of several ice shelves in the northern part of the peninsula over the last two decades.

Between 1900 and 2008 the global ocean warmed by about 0.6 °C, but the western boundary currents (the Gulf Stream, Kuroshio, the Brazil, Aguhlas, and East Australian currents) warmed by 1.0–1.4 °C. This is attributed to a poleward shift in the mid-latitude extensions of these currents. The role of winds is unclear, as in the Southern Hemisphere the westerlies strengthened, but not in the Northern Hemisphere. Nevertheless, these warmer currents will affect the ocean–atmosphere heat transfers and the development of cyclones. During the last 3–4 decades, the frequency of El Niño events has increased, while La Niña events decreased. This has been attributed by some scientists to global warming, but the case is not yet established. In the last three decades there has been an expansion of the Tropics in the Northern Hemisphere. Various metrics show an average expansion between 1979 and 2009 of 0.33°/decade. Allen *et al.* attribute this shift to the heating effects of black carbon aerosol and tropospheric ozone, especially in the warm season.

The relative frequency of cold spells has decreased and the frequency of hot days has increased over most land areas in the late twentieth century. In Europe the frequency of hot days has tripled since 1880. In the United States there has been an increase of about 12 days in extreme (85th percentile) apparent temperatures in the 60 years since 1949. In the Colorado alpine zone, the area classified as tundra according to the Köppen classification shrank by 73 percent between 1901–1930 and 1987–2006 as a result of summer warming, according to Diaz and Eischeid.

Drought areas have increased, especially in the Tropics and Subtropics. There has been widespread drying over Africa, East and South Asia, eastern Australia, southern Europe, Alaska, and northern Canada from 1950 to 2008; most of this drying is due to recent warming. Since the 1970s the area characterized as very dry by the Palmer Drought Severity Index has more than doubled. At the same time it has become wetter in the United States and most of Western Australia. The global percentage of dry areas has increased by about 1.7 percent of global land area per decade from 1950 to 2008. Nevertheless, the frequency of heavy precipitation events has also increased in most land areas. This is mainly observed in regions and seasons where there is a strong moisture transport from the ocean to the land, rather than in summer when there is interplay between soil moisture and convection. Increases in the frequency and intensity of North Atlantic hurricanes have also been attributed to global warming, but the case for this is still being debated.

SUMMARY

There have been several major ice ages in Earth history. Snowball Earth conditions occurred twice in late Proterozoic time. Since 500 Ma there were glaciations in the Ordovician, Permo-Carboniferous, and late Cenozoic, and a major ice-free climate in the Cretaceous associated with high CO_2 levels in the atmosphere. A major driver of climate change on the geological time scale has been plate tectonics and associated continental drift, mountain building, and volcanic activity. In the early Cenozoic around 65–50 Ma global climate was warm, but then a cooling trend began that led to ice build-up in East Antarctica around 33 Ma. The continents were in their present positions by 17–15 Ma. Perennial sea ice formed in the Arctic about 14 Ma, but there was no major land ice cover in the Northern Hemisphere until around 3 Ma. The Quaternary period that began at 2.6 Ma witnessed numerous glacial cycles that are identified in the marine isotopic record of $\delta^{18}O$. Until 800 000 years ago there was a dominant obliquity signature of 41 000 years, that then switched to 100 000 years, although the orbital eccentricity effect on solar radiation is minor. For recent glacial cycles, glacial conditions took up 90 percent of the time. The last interglacial (the Eemian) peaked around 125 ka, with temperatures higher than in the twentieth century. Cooling began around 115 ka and there was a full glacial by 75 ka. Conditions moderated and then the Last Glacial Maximum occurred around 21 ka. Global temperatures fell about 6 °C and sea level dropped 130 m. The glacial cycle was marked by rapid c.1500-year climate fluctuations known as Dansgaard–Oeschger oscillations, related to changes in North Atlantic circulation. The Younger Dryas cold event (12.8–11.5 ka) is attributed to a massive influx of freshwater into the North Atlantic from glacial Lake Agassiz. In Africa cold, arid phases marked the glacial conditions.

The post-glacial Holocene that began at 11.7 ka was less variable than the Pleistocene. A further orbital factor affecting climate is the precession of the equinox; currently perihelion is in January, but 10 000 years ago it was in July, making northern summers hotter and wetter. Savanna covered the Sahara from 9500 to 6500 BP. The Holocene thermal maximum was from 8000 to 5500 BP.

Around AD 950–1250 there was a Medieval Warm Period in Europe and the North Atlantic, followed by the Little Ice Age (AD 1400–1700 in the Northern Hemisphere), with glacier advances in Europe peaking in the mid to late nineteenth century. Volcanic eruptions in the late thirteenth century may have triggered the Little Ice Age.

The Anthropocene – marked by pronounced human effects on global climate – is dated to about AD 1800, the start of the Industrial Revolution. Greenhouse gas effects were first noted in 1896. Water vapor accounts for about half of the greenhouse effect. CO_2 measurements began in 1957 with earlier data obtained from ice cores. Since 1800, CO_2 levels have increased by 40 percent from 280 to 391 ppm, and methane rose 2.5 times. The total radiative forcing is about 1.6 W m^{-2}. Atmospheric aerosols led to a global dimming over the 1950s–1980s, but this has now reversed, except for India and China. Mean global temperature rose by 0.7 °C in the twentieth century. In the Arctic the warming was twice as great, partly as a result of ice-albedo feedbacks and, in the 2000s, Arctic sea ice shrank dramatically in September. Almost worldwide, glaciers have retreated greatly since the mid twentieth century. The frequency of hot days and heat waves has increased. Also, the frequency of heavy precipitation increased in most land areas.

QUESTIONS

1 Compare the proxy records available to analyze centennial changes in climate over the last 10 000 years.
2 Characterize the main changes in global climate over the last two million years.
3 Compare climate changes in northern Europe and North Africa over the interval from 25 ka to 5 ka.
4 Identify the three orbital effects by name and state their periodicities.
5 Compare the characteristics of $\delta^{18}O$ in marine sediments and in ice cores and explain their contrasting signatures.
6 Characterize the climates of the Medieval Warm Period and the Little Ice Age.
7 Compare the role of natural variability and anthropogenic forcing on climate since AD 1800.
8 Explain how the greenhouse effect operates.
9 Consider the impacts of global warming on the environment.
10 Summarize the factors that account for the total radiative forcing of the Earth and state their relative importance.

11 Future climate

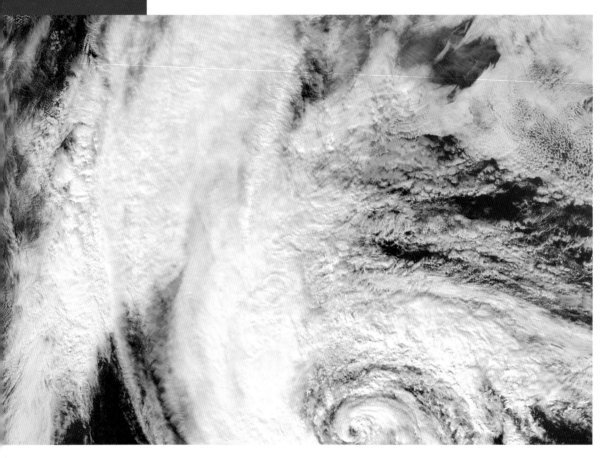

Hurricane Sandy off the east coast of the United States on 28 October 2012.

AS WE noted in the last chapter, the last 30 years have seen significant global warming, with nearly universal retreat of mountain glaciers, and major losses of Arctic sea ice cover in the 2000s, especially in late summer. Ice shelves have collapsed off northern Ellesmere Island, Canada, and in the Antarctic Peninsula. The Intergovernmental Panel on Climate Change (IPCC) issued its Fourth Assessment Report in 2007, concluding that increased concentration of greenhouse gases (GHGs) was almost certainly the cause. The report also projected global climatic conditions by AD 2100 using the results of simulations with coupled global climate models.

We begin with a brief review of global climate models, and then consider the major projected changes during this century, their impacts, and finally related economic and socio-political issues.

11.1 Global climate models

A large number of model projections of GHG concentrations in AD 2100 have been made using different scenarios of population growth and economic activity, These scenarios have been applied in *global climate models* since the third IPCC report in 2001. Global climate models are coupled atmosphere and ocean general circulation models (GCMs) that treat mathematically almost all climatic processes at high temporal and spatial resolutions.

The basic equations for an atmospheric GCM are those of air motion, continuity of mass and of water substance, the thermodynamic equation, and the hydrostatic equation relating gravity and vertical pressure gradient. The equations are solved iteratively at about five-minute intervals. Typical atmospheric GCM grids have 1–5 degree latitude spacing and represent topography, surface vegetation, and land cover as boundary conditions (see Figure 11.1). They have numerous

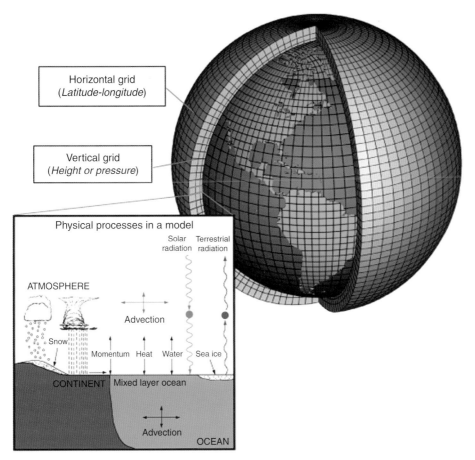

Figure 11.1 Schematic of the grids used in atmospheric models and the physical processes included (source: NOAA).

Box 11B.1 The Climate Modeling Intercomparison Project (CMIP)

In September 2008, 20 climate modeling groups from around the world, the WCRP's Working Group on Coupled Modelling, with the International Geosphere Biosphere Programme's Analysis, Integration, and Modeling of the Earth System (AIM) project, agreed to promote a new set of coordinated climate model experiments. These experiments comprise the fifth phase of the *Coupled Model Intercomparison Project* (CMIP5). The project will provide a multi-model context to (1) assess the mechanisms responsible for model differences in poorly understood feedbacks associated with the carbon cycle and with clouds, (2) examine climate "predictability" and explore the ability of models to predict climate on decadal time scales, and (3) determine why similarly forced models produce a range of responses. Results of these experiments are incorporated in the IPCC's Fifth Assessment Report, issued in September 2013. The following suite of experiments is scheduled: (1) Decadal hindcasts and prediction simulations, (2) "long-term" simulations, and (3) "atmosphere-only" (prescribed sea-surface temperature [SST]) simulations for especially computationally demanding models. Hindcasts are "forecasts" that are made for past decades where observational data are available for comparison with the model simulations. Detailed information about CMIP5 and links to publications are available at http://cmip-pcmdi.llnl.gov/cmip5

levels in the vertical that extend into the upper atmosphere, as do ocean circulation models with depth.

Models of the atmosphere were first developed in the 1950s, with first-generation GCMs available in the 1970s. Gradually, these atmospheric GCMs were coupled first with a slab ocean and then with ocean GCMs.

Later, submodels of land processes and sea ice were added. CO_2-doubling experiments were first carried out in the 1980s. Here we will consider the outcomes of a number of the IPCC experiments performed in 2005–2006. It should be noted that the mean values of an ensemble of model results generally agree well with observations of twentieth-century climate change, while the results of individual models may show considerable scatter. Hence, it is now usual to average the results of many models in an ensemble. A major activity in the modeling community in recent years has been the Climate Modeling Intercomparison Project (CMIP). This is summarized in Box 11B.1.

11.2 Projected changes

An important concept is the equilibrium *climate sensitivity*, which refers to the expected global warming for a sustained doubling of carbon dioxide concentration to 600 ppm. This value is estimated to be between 2 °C and 4.5 °C, with a most

Table 11.1 The four families of scenarios developed in 2000 and used in the Fourth Assessment Report of the IPCC

	Economic focus	Environmental focus
Globalization (homogeneity)	A1: Rapid economic growth	B1: Global environmental sustainability
Regionalization (heterogeneity)	A2: Regionally oriented economic development	B2: Local environmental sustainability

Note: For the Fifth Assessment Report (September 2013), these scenarios will be replaced by Representative Concentration Pathways (RCPs). RCP2.6, RCP4.5, RCP6, and RCP8.5 are named after a possible range of radiative forcing values in the year 2100 (2.6, 4.5, 6.0, and 8.5 W m^{-2}, respectively).

likely value of 3.0 °C. The radiative forcing of 3.7 W m^{-2} due to CO_2 doubling, without any feedbacks, is readily determined to result in a warming of 1 °C; the additional warming is caused by mainly positive feedbacks involving water vapor, ice-albedo, clouds, aerosols, and temperature lapse rate effects.

Based on different scenarios (see Table 11.1) of global population, economic activity, and energy consumption, concentrations of atmospheric CO_2 are projected to rise from 390 ppm in 2011 to between 535 and 983 ppm by AD 2100. Methane, currently 1825 ppb, is projected to range between a small decrease to 1460 ppb, and an increase to 3390 ppb. The uncertainty in these projections is also a result of the problems of projecting changes in the global vegetation cover. A further uncertainty is the thawing of permafrost releasing additional methane. Nevertheless, using these scenarios, and changes in aerosol loading, the models are then run to simulate changes in climate throughout the world during the twenty-first century.

Global temperatures by 2090–2099 are projected to increase, relative to average temperature for 1980–1999, by between 1.8 °C (scenario B1 – rapid change in economic structures toward a service and information economy, with reductions in material intensity and the introduction of clean and resource-efficient technologies) and 4.0 °C (scenario A1FI – very rapid economic growth, intensive use of fossil fuels, global population that peaks mid-century and declines thereafter). Terrestrial temperatures in the Arctic will increase between 3 and 8 °C in the same time frame. However, in winter the Arctic warming will be four times the global average due to polar amplification, as discussed in Section 10.4. Figure 11.2 shows a multimodel average projection of temperature, precipitation, and pressure for 2080–2099, illustrating the large amount of warming and precipitation changes that are projected. The latest multi-model simulations project a larger global warming of between 1.4 °C and 3.0 °C by 2050 relative to 1961–1990, using the

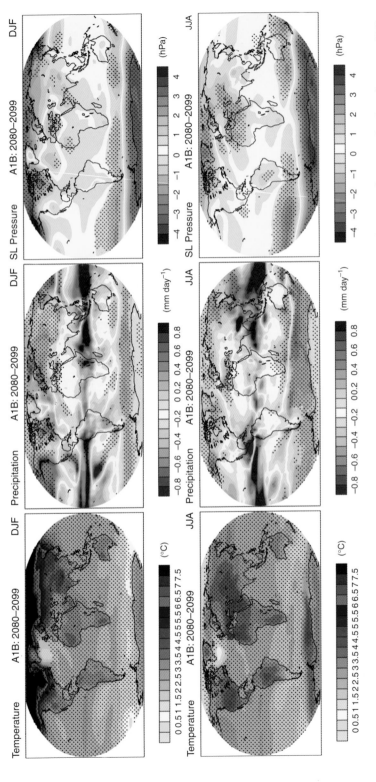

Figure 11.2 Projected changes for 2080–2099, relative to 1980–1999, in temperature, precipitation, and pressure for winter (DJF) and summer (JJA). The projections are a multi-model average for the A1B scenario for the climate models used in the IPCC Fourth Assessment Report (source: Solomon *et al.* 2007).

A1B mid-range forcing scenario without mitigation. The principal uncertainties in these estimates involve the value for equilibrium climate sensitivity, the uptake of heat by the oceans, and the history of aerosol loading in the atmosphere.

The 22 IPCC coupled climate models project a variety of changes in rainfall according to region, although the increases in atmospheric vapor content due to higher temperatures should lead to more rainfall. In particular, the models project an increase in rain intensity, particularly in tropical and high-latitude regions that experience overall increases in precipitation. Annual average precipitation will increase over most of northern Europe, Canada, the northeastern United States, tropical and eastern Africa, the northern Pacific, the Arctic, and Antarctica, as well as northern Asia and the Tibetan Plateau in winter. Annual average precipitation will decrease in most of the Mediterranean, northern Africa, northern Sahara, Central America, the American southwest, the southern Andes, as well as south-western Australia during winter. By the 2060s there will be severe drought over most of Africa, southern Europe and the Middle East, most of the Americas (except Alaska and northern Canada, Uruguay, and northeastern Argentina), Australia, and Southeast Asia, according to A. Dai.

Mid-latitude storm tracks are projected to shift poleward, but there will be a reduced frequency of storms, both changes reflecting the weakening of the Equator–Pole temperature gradient due to enhanced high-latitude warming. Tropical storms and cyclones are likely to become more intense, with stronger peak winds, and produce increased rainfall over some areas due to warming sea surfaces. However, there is no consensus on changes in tropical storm frequency.

General circulation model experiments can also be used to assess projected trends in circulation patterns. For Europe, an ECHAM5 climate model integration with the A1B emission scenario for 1860–2100 projects a linear trend of $+12.2$ days per 240 years for westerly types and -7.0 days per 240 years for easterly types. Hence, the circulation will become more zonal.

11.3 Impacts

Societal and environmental impacts of climate change are numerous. A major concern is that higher temperatures are expected to raise mean sea level. Three factors are involved in the process of *sea-level rise*: thermal expansion of the ocean (the steric effect), melting of mountain glaciers and small ice caps, and increased melting or calving of major outlet glaciers of the Greenland and Antarctic ice sheets. During 1961–2003 these three factors contributed 0.42, 0.5, and 0.19 mm yr^{-1} to sea-level rise, respectively. Thermal expansion of the oceans raises sea level 0.2–0.6 m per degree Celsius of globally averaged surface warming.

Since the 2007 IPCC report appeared, there have been new observations and calculations of accelerated retreat of the ice sheets. The recent Arctic Monitoring and Assessment Program (AMAP) projections for sea-level rise by AD 2100 are in the range 0.9–1.6 m, about double the previous estimates, as a result of accelerating ice loss from Greenland and West Antarctica. This rise will have severe impacts on coastal populations such as those in Bangladesh, southern Florida, Pacific atolls, and other low-lying coastlines. Model calculations of the amount of warming needed to melt the Greenland ice sheet completely, and raise sea level by about 7 m, indicate a threshold temperature of only about 2 °C above pre-industrial levels, according to Robinson *et al.* However, the time scale involved in this would likely be several thousands of years.

Spring snowmelt is occurring about eight days earlier in northern Alaska than in the 1960s, and several weeks earlier in places such as the North American cordilleras. In the Pacific Northwest, Sierra Nevada, and Rocky Mountains, this is advancing the date of peak runoff by up to a month and reducing the summer flow in rivers. This effect will intensify over this century, leading to serious water shortages in summer, affecting domestic consumption, hydroelectric power generation, and irrigation.

The warming in the Arctic, through the advection of warmer waters via the North Atlantic and Bering Strait, and more frequent incursions of warm air, is projected to lead to an ice-free Arctic Ocean in summer by 2030–2040, or possibly sooner. This will have enormous economic and environmental consequences. The open Arctic will provide shorter sea routes between East Asia and North America and it will permit easier oil and gas exploration of the seabed. At the same time it will largely eliminate ice-based ecosystems – seals, narwhals, and polar bears – and radically change the life styles of indigenous peoples in Alaska, Arctic Canada, northern Scandinavia, and northern Russia.

High-latitude land areas are underlain by great thicknesses of permanently frozen ground or *permafrost*. This may be frozen bedrock or it may consist of ice-rich organic matter. In northern Alaska and Siberia the permafrost is continuous, with a thickness of 500–600 m and a maximum of 1000 m in north central Siberia. The area of continuous permafrost in the Northern Hemisphere is 17 million km^2, while to the south there is a further 17 million km^2 of thinner, discontinuous and sporadic permafrost. A map of Northern Hemisphere permafrost can be accessed at http://nsidc.org/data/ggd318.html

Seasonally, the surface layer of the permafrost thaws to about 50 cm depth in the north and 3–4 m in the south. The seasonal depth of the active layer below the surface of the permafrost across the Arctic and Subarctic is likely to increase progressively as Arctic temperatures continue to rise. Thawing of ice-rich permafrost is giving rise to local subsidence of the ground, termed thermokarst, forming depressions and lakes. This subsidence creates major problems for infrastructure – roads, railways, pipelines, airstrips, and buildings – unless special engineering techniques are employed. Thin permafrost near its southern margins and lower altitudinal limits today will disappear in this century. The duration of seasonal ground freezing will be shortened, with implications for agriculture.

A major long-term problem caused by permafrost thaw is the release of methane – a potent greenhouse gas (GHG) – from the thawed organic matter. It is estimated that the total volume of methane currently locked up in permafrost is twice that contained in the atmosphere. Hence, there is a substantial potential for further temperature rise and a significant feedback effect on global warming. This contribution to global carbon also does not take into account the methane content held in subsea permafrost off the north coasts of Alaska and Siberia, whose extent and thickness is very poorly known.

Global warming is leading to an increased frequency of extreme warm events. Already the numbers of hot days, hot nights, and heat waves have gone up and the frequency of cold events has decreased. In August 2003, a heat wave in Europe resulted in excess mortality of the order of 40 000 deaths (see Section 12.1). In July 2011, most of the central and eastern United States sweltered under temperatures of around 38 °C, with heat indices (taking humidity into account) of 41–48 °C.

Higher temperatures are already being reflected in the shift of plant species and animal species. In 2012 the US Department of Agriculture, Agricultural Research Service, issued a new map of plant hardiness zones for the United States that takes account of warmer climatic conditions. The original 1990 map used data for 1974–1986, whereas the new map by Daly *et al.* is based on 1976–2005 data. A warming of at least 5.6 °C in the extreme winter minimum temperature has occurred over large parts of the central and eastern United States, as well as the Pacific Northwest. In European mountains the alpine zone is seeing a shift toward warmer plant species with rising temperatures. In another example, penguins in the Antarctic Peninsula are being forced southward due to higher temperatures. The changes in temperature and rainfall will impact agriculture in complex ways. Higher CO_2 levels will tend to augment plant growth. Higher temperatures will tend to increase (decrease) crop yields in high (low) latitudes.

Increased rainfall intensity may damage crops, while lower rainfall amounts will lead to lower yields or harvest failure, leading to famine. The distributions of insect pests and diseases affecting plants, animals, and humans will also change in response to climate change. The environment, a host, and a pathogen, all of which are affected by temperature, humidity, and soil moisture, determine plant diseases. Much research in this area is needed to determine the potential impacts.

11.4 Economic and socio-political issues

Widespread concern over the risks to society posed by continuing global warming have led to numerous efforts to address these issues by nations, state governments, non-governmental organizations, corporations, cities, and individual citizens. The principal risks are identified to be the effects on human health, agriculture, plant and animal species, and insect pests and diseases caused by rising temperatures and changing precipitation conditions; rising sea levels threatening low-lying coastlines; water supplies affected by shrinking glaciers and snow packs; and major climate changes in high latitudes of the Northern Hemisphere resulting from the loss of summer sea ice cover in the Arctic Ocean.

Because the primary cause of global warming is the continuing increase in atmospheric concentrations of GHGs, the main focus of attention has been on ways to cut emissions of GHGs. Attention is also being given to technological means to capture and sequester carbon. It is estimated that a warming of 2 °C, of which 0.8 °C has already occurred, will accompany a CO_2 concentration of about 450–500 ppm – the level projected for about 2060. A warming of 2 °C is considered to be the threshold of dangerous human consequences.

In 1997 the Kyoto Protocol was initially adopted by nations operating under the United Nations Framework Convention on Climate Change (UNFCCC) and entered into force in 2005. The ultimate objective of the UNFCCC is to stabilize GHG concentrations in the atmosphere at a level that prevents dangerous anthropogenic interference with the climate system. The main aim of the Protocol is to contain emissions of the main human-emitted GHGs in ways that reflect underlying national differences in such emissions, wealth, and capacity to make the reductions. Currently, 190 states have signed and ratified the protocol; the United States has signed but not ratified it, and Canada withdrew in 2011. The protocol expired in December 2012 and both the Copenhagen conference in 2011 and the Rio + 20 summit in 2012 failed to agree to a successor, leaving the planet and its population in peril.

There are two strategies to combat global warming – mitigation and adaptation. Mitigation involves human intervention to reduce the sources or enhance the sinks of GHGs, thereby lessening the radiative forcing that leads to temperature rise. Adaptation seeks to reduce the vulnerability of natural and human systems to the effects of climate change.

The burden of mitigation falls on developed economies because they have the largest share of GHG emissions. For example, the United States has been responsible for 29 percent of total global GHG emissions since 1850. Currently, China accounts for 24 percent of the world total and the United States for 18 percent. In

Box 11B.2 **Geoengineering**

Geoengineering refers to deliberate large-scale intervention in the Earth's climate system in order to offset global warming.

There are two broad approaches: (1) CO_2 removal techniques which address the root cause of climate change by removing GHGs from the atmosphere; and (2) solar radiation management techniques which attempt to offset effects of increased GHG concentrations by reducing the absorption of solar radiation.

Techniques of carbon removal include:

- creating biochar (anaerobic charcoal) and mixing it with soil;
- bioenergy with carbon capture and storage to sequester carbon and simultaneously provide energy;
- carbon air capture to remove CO_2 from ambient air;
- ocean nourishment, including iron fertilization of the oceans.

Solar radiation management techniques include:

- surface-based land or ocean albedo modification, e.g., using pale-colored roofing and paving materials;
- troposphere-based, e.g., cloud whitening using fine seawater spray to whiten clouds and thus increase cloud reflectivity;
- upper atmosphere-based, e.g., stratospheric aerosols. Creating reflective aerosols, such as sulfur aerosols, aluminum oxide particles;
- space-based, e.g., space sunshade obstructing solar radiation with space-based mirrors, asteroid dust, etc.

2006 Elsevier launched the *International Journal of Greenhouse Gas Control*, which has papers on CO_2 capture and storage technology. Forest clearance and burning is the source of about 15–20 percent of global CO_2 emissions. One of the related mitigation approaches is the United Nations program for the Reduction of Emissions from Deforestation and Degradation (REDD) in developing countries. International funding is being provided to maintain tropical rainforests (see www. un-redd.org)

The capture and sequestration of CO_2, at present, is mainly at the research and development stage, but there are small-scale operational projects where CO_2 is stored deep underground. Further into the future are *geoengineering* projects that involve modifications to the incoming solar radiation by injecting absorbing aerosols into the stratosphere, for example, or by increasing the reflectivity of the ground surface (Box 11B.2). Although many of these techniques hold promise, most of them are very controversial due to the possibility of unexpected adverse effects. Much more research in this field is called for before any experiments could be undertaken.

Adaptation is especially important for developing countries since they are predicted to bear the brunt of the effects of climate change. Examples of adaptation include building defensive structures against rising sea level, or relocating vulnerable populations, and changing crops planted to adapt to climate changes. Local responses are important in the case of health hazards from heat waves. Strategies include emergency warnings targeted at populations at particular risk, building cold emergency shelters or cold rooms in apartment buildings, and funding for air conditioning for low-income people.

Since 2009 much attention has been given to the provision of funding to undeveloped countries for adaptation and also mitigation measures. This has been prompted by the increasing frequency of climate-related disasters. So-called Fast Start Finance totaling US $30 billion has been pledged by developed countries for 2010–2012, with larger sums in subsequent years. Issues that arise in these plans concern: (1) Who pays, by what mechanisms, and how much? (2) To whom is the money allocated and how are projects prioritized? (3) Who controls the finances? A status report is available at www.climatefundsupdate.org

SUMMARY

General circulation models were first developed in the 1970s and used for CO_2-doubling experiments in the 1980s. The estimated equilibrium sensitivity of global climate for CO_2-doubling is estimated to be 3 °C. Atmospheric CO_2 concentration is projected to reach between 535 and 983 ppm by AD 2100, and Arctic land temperatures will rise between 3 °C and 8 °C. Increased vapor content will increase precipitation in tropical and high-latitude regions.

Global sea level will rise between 0.9 and 1.6 m by AD 2100 as a result of ocean thermal expansion and land ice loss. By AD 2030–2040 it is projected that the Arctic Ocean will be ice-free in summer. On land, snowmelt and peak runoff will occur earlier and the upper permafrost layer will thaw. Extreme warm events will increase in frequency and vegetation, agricultural crops, and insect pests and diseases will respond in complex ways to changes in temperature and rainfall.

Socio-political and economic concerns have focused on means to reduce greenhouse gas emissions, on carbon capture and sequestration, and on reducing rainforest clearance. The Kyoto Protocol was adopted in 1997. The goal is to stabilize greenhouse gas concentrations at a level that prevents dangerous impacts on the climate system, such as a warming exceeding 2 °C. Strategies to respond to climate warming include mitigation (reducing emissions, carbon capture, or geoengineering) or adaptation (e.g., changing crops planted). Undeveloped countries will be most impacted and efforts to provide funding assistance from the developed world are underway.

End Note

The IPCC has just released its Fifth Assessment Report (AR5). Highlights are as follows: Global mean temperature increased by 0.85 °C during 1880–2012 and twice as much in the Arctic. A slowing in the rate of warming since 1998 to 0.05 °C/ decade is probably related to increased uptake of heat in the deep ocean, reduced radiative forcing, and natural variability. Projections for 2081–2100 are for a rise in global mean temperature greater than 1.5 °C for all RCP scenarios and greater in the Arctic. The warming is almost certainly attributable to the emission of greenhouse gases through human activity; CO_2 concentrations will double by 2050. The area of Arctic sea ice is projected to decline and the Arctic will likely become ice-free in summer before 2050 for RCP 8.5 (W m^{-2} radiative forcing). Glaciers will continue to shrink and the Greenland and Antarctic ice sheets will lose mass at an accelerating rate due to enhanced surface melt and iceberg calving. Sea levels will rise by up to 80 cm (RCP 6.0) by 2100, creating massive problems for coastal cities. Snow cover and near-surface permafrost area will shrink. Extreme precipitation events will increase in intensity and frequency over mid-latitude land and in the wet tropics.

QUESTIONS

1 What are the major uncertainties in model projections of twenty-first-century climate?
2 Why do projected CO_2 levels for 2100 show such a wide range?
3 What are the major projected impacts of global warming in the twenty-first century, and which do you consider to be the most serious?
4 What are the main contributors to rising sea level and what are their relative contributions?
5 Compare the effects of rising sea level in Florida and Bangladesh.
6 What will be the consequences of an open Arctic in summer?
7 Give examples of mitigation and adaptation strategies to combat climate change.
8 Review and compare the Montreal and Kyoto Protocols.
9 Critically evaluate the proposed techniques for geoengineering to address climate change.

12 Applied climatology

Turbines at a wind farm.

IN THIS chapter we examine various examples of the application of climatic information. We begin by examining climatic extremes and disasters. Then we examine climate and soils, and climate and agriculture. This is followed by a consideration of water resources, renewable energy, and climate and transportation. The chapter concludes with a brief summary of insurance and climate-related disasters and a section on climate forecasts and services.

12.1 Climatic extremes and disasters

Much attention has been given to weather extremes – severe thunderstorms, tornadoes, windstorms and tropical cyclones. These events typically last from a few hours to a few days. However, *climatic extremes* have time scales from a month to decades, with long-lasting consequences. They include persistent intense rains giving rise to floods, protracted droughts, heat waves, and freezes.

For the United States a Climate Extremes Index (CEI) has been developed by NOAA, which is based on an aggregate set of conventional climate extreme indicators. These include the following types of data:

- monthly maximum and minimum temperature;
- daily precipitation;
- monthly Palmer Drought Severity Index (PDSI);
- the wind velocity in land-falling tropical storm and hurricanes.

Temperature and precipitation data are taken from 1220 stations in the US Historical Climatology Network (USHCN) covering the contiguous United States. The values are based on the percentage of the area of the United States with much below/much above normal maximum and minimum temperatures and, similarly, of severe drought or extreme moisture surplus. The details may be found at www.ncdc.noaa.gov/extremes/cei

Between 1910 and 2011 the CEI ranged from 9 to 40. It was lowest over 1942–1980 and has subsequently been consistently higher.

The IPCC states that since 1950 the number of *heat waves* has increased. The extent of regions affected by droughts has also increased as precipitation over land has marginally decreased while evaporation has increased due to warmer conditions. Generally, the number of heavy daily precipitation events that lead to flooding have increased, but not everywhere. For the period 2000–2011, Coumou and Rahmstorf list 18 events that were record breaking and estimate that climatic warming has increased the number of new global-mean temperature records expected in the last decade from 0.1 to 2.8.

We will now consider some examples of major extreme events. The European heat wave of August 2003 saw the highest temperatures since at least AD 1540 and caused an estimated 40 000 fatalities. The heat wave was concentrated in France, Germany, and northern Italy. Temperatures topped 40 °C and most inhabitants lacked air conditioning. The primary cause was a persistent anticyclone over western Europe and warm air advection. It is uncertain whether or not the record temperatures can be attributed to global warming, but a human contribution to the risk of such an extreme event has been shown to be statistically very likely. In July–August 2010 western Russia experienced a prolonged heat wave and this gave rise to extensive wildfires that affected 1.5 million ha. It was the hottest summer recorded in Russia and many days topped 35 °C. In Moscow the July temperature was 18 °C above average. It is estimated that in Russia about 55 000 fatalities resulted from the heat and smoke from the fires. The primary cause of the event was persistent atmospheric blocking, amplified by a positive feedback due to below-normal soil moisture conditions. There is an 80 percent probability that the 2010 July record heat in Moscow would not have occurred without climate warming, according to Coumon and Rahmstorf.

In summer 1993 there was extensive flooding in the Mississippi River basin. The problem began in autumn 1992 when the soil became saturated as a result of cooler conditions reducing evaporation. Winter rains and spring snowmelt added to the runoff into streams. January–July precipitation was twice the average. Then a persistent flow of warm, moist air running from the Gulf of Mexico to the Midwest set up numerous thunderstorms during June to August, with between 200 and 350 percent of the average falling on parts of the basin. Peak discharge in the Mississippi and Missouri rivers substantially exceeded the 100-year flood level. In total, 50 000 homes were destroyed, there were 48 deaths, and losses totaling US $10 billion. There was also a comparable flood in winter 1937 that devastated Cincinnati and Louisville on the Ohio River, and yet another in 2011. Both are estimated to have been events with a return period of 200 years.

In late July 2010, 300–400 mm of rain fell in the Khyber Pakhtunkhwa, Sindh, Punjab, and Baluchistan regions of Pakistan, leading to severe flooding in the Indus basin. Up to one-fifth of the land area of Pakistan was under water. The death toll approached 2000, but 20 million people were affected by the flooding, losing homes and crops. Lau and Kim suggest a teleconnection between the blocking pattern that caused the Russian heat wave, setting up a downstream trough. Strong southeasterly flow along the Himalayan foothills advected abundant moisture from the Bay of Bengal over Pakistan.

Drought is a recurrent phenomenon in semi-arid environments such as the western United States, the Sahel, the Horn of Africa, and Australia. In the Midwestern United States the Dust Bowl years of the 1930s are well-known examples (see Box 9B.2). The period 1930–1936 saw recurrent droughts that, combined with bad farming practices, allowed winds to remove the topsoil, creating massive dust storms. Tens of thousands of farmers were forced to migrate, as described in John Steinbeck's famous novel *The Grapes of Wrath*. In Medieval times in the western United States there were four mega-droughts, each of which lasted for several decades. These were centered around AD 936, 1034, 1150, and 1253 during the Medieval Climate Anomaly, according to Cook *et al*.

In the Sahel of West Africa, a shift of rainfall regime in the 1970s led to dry conditions that continue up to the present. This contrasts strikingly with the wet interval of 1950–1970 that allowed cattle numbers and population to increase sharply. North of 15° N, mean rainfall during 1970–1984 was 50 percent or even less than during 1950–1959. In contrast, from 1900 to 1950 conditions were variable with no long wet or dry spells. The cause of the regime shift has been attributed to an expansion of the arid core of the Sahara and also to a change in pattern in sea-surface temperatures (SSTs) in the North Atlantic.

In December 2010 three unprecedented freeze events in Florida severely impacted the sugar cane crop. Freezes occurred on 7–8 December (temperatures below − 1 °C) and on 14–15 and 27–28 December (temperatures below − 2 °C). Cold air outbreaks led to record-breaking low temperatures in east-central Florida. In late January–February 2012 a protracted severe freeze and heavy snowfall affected eastern and southeastern Europe as Arctic air moved southeastwards from northern Russia. Temperatures fell to below − 30 °C, and ice blocked the Danube River. Over 550 deaths were attributed to the extreme cold.

A recent survey by Peterson *et al.* of a number of climatic extremes of temperature and precipitation that occurred in 2011 attempts to assess the probability of their occurrence and their attribution. While it is emphasized that not all extremes can be blamed on human influences on atmospheric composition, long-term warming trends are affecting the frequency and/or intensity of climatic extremes. For example, the Texas heat wave and drought of May–August 2011 were the most severe on record (dating back to 1895) and are attributed in part to the increase in greenhouse gas (GHG) levels over the last 50 years. Annual reports of this type are planned for the future to accompany the annual "State of the climate" reports published as supplements to the *Bulletin of the American Meteorological Society*.

12.2 Climatic aspects of vegetation and soils

A global view of vegetation is shown in Figure 12.1 based on the Normalized Difference Vegetation Index (NDVI). Clearly visible are the deserts and tropical rainforests. The climatic characteristics of the major vegetation zones are presented in Chapter 9. Also, in Section 2.2e we showed the relationship of major vegetation categories to Budyko's radiational index of dryness. However, we did not consider the association of climate and soil types. Here we will briefly examine this relationship.

Soils comprise weathered rock and organic matter that is differentiated into layers, or horizons, of variable depth. Climate is the key control on a global scale, although parent material, slope, drainage, and time of development may modify its influence. A first-order effect is determined by whether P > E or P < E. In the former case, water infiltrates downward and leaches minerals from the soil. Locally, waterlogging may occur, leading to the development of gray-blue gleys, which can occur in almost all climatic zones. These wetland soils occupy almost

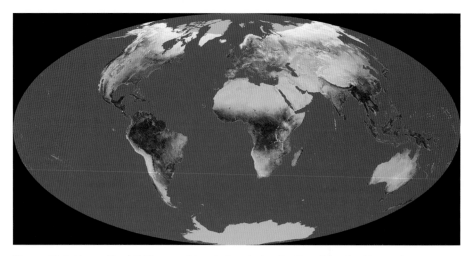

Figure 12.1 Normalized Difference Vegetation Index for the globe for November 1– December 1, 2007 (source: NASA Earth Observatory image produced by Reto Stockli and Jesse Allen, using data provided by the MODIS Land Science Team, http://earthobservatory.nasa.gov/IOTD/view.php?id=8622).

5 percent of the land surface. Where P < E, salts may be deposited on the soil surface.

An important soil variable is its degree of acidity, measured by pH (which is the negative logarithm to base 10 of the concentration of hydrogen ions). It has a range from 0 to 14, with a value of 7 being neutral; lower values are acidic and higher values basic. Rainfall generally has a pH of ∼ 5.7. Basic soils have a high proportion of calcium, magnesium, potassium, and sodium cations (positively charged ions).

Soil temperature and soil moisture are two very important climatic elements. Soil temperature is measured at standard depths by soil thermometers at many stations around the world. Soil moisture is less easily measured and the representativeness of the data is uncertain. However, a global soil moisture data set for 1978–2010 has recently been released by the European Space Agency (ESA). It developed the information using C-band scatterometers and passive microwave data from a variety of satellite sensors. The data can be accessed at www.esa-soilmoisture-cci.org/node/127

The US Department of Agriculture soil classification developed in 1978 identifies the following types:

- Alfisols form in semi-arid to humid areas that have a clay- and nutrient-enriched subsoil. They form under forest vegetation where the parent material has undergone significant weathering. The most distinguishing characteristic is the deposition of clay in the B horizon. They occupy 10 percent of the land surface.

- Andosols are black soils formed on volcanic tuff and ash. They have a wide distribution and occupy about 2.5 percent of the land surface.
- Aridosols develop in dry steppe environments. Their main characteristic is poor and shallow soil horizon development. Because of low rainfall and high temperatures, soil water tends to migrate upward. This leads to the deposition of salts carried by the water at or near the ground surface due to evaporation. A subgroup that has sodium deposits in the upper horizon is known as solonetz or alkali soils. They occupy about 12 percent of the land surface.
- Entisols are soils of recent origin developed in unconsolidated parent material with usually no genetic horizons except an A horizon. They occupy 18 percent of the land surface.
- Gelisols commonly have a dark organic surface layer and mineral layers underlain by permafrost. The alternate thawing and freezing of ice layers results in special features such as frost-heaving and deformed landscapes. The slow decomposition of the organic matter due to low temperatures results in the formation of a surface peat layer. They occupy about 9 percent of the land surface.
- Histosols are organic peaty soils that form in areas of poor drainage. They have decomposed organic materials derived from sedges, grasses, leaves, hydrophytic plants, and woody materials. They occupy about 1 percent of the land surface.
- Inceptisols are soils that exhibit minimal horizon development. They are found on young surfaces or steep slopes. They occupy about 15 percent of the land surface.
- Mollisols are the soils of grassland ecosystems such as the prairies. They are characterized by a thick, dark surface horizon. An important subgroup is the chernozem or black earth. They occupy 7 percent of the land surface.
- Oxisols are very highly weathered soils found primarily in the intertropical regions. They contain few weatherable minerals and are often rich in iron and aluminum oxides. They occupy about 7.5 percent of the land surface.
- Spodosols are acid soils characterized by an ash-gray subsurface underlain by an accumulation of humus that contains iron and aluminum. They occur under coniferous forest in cool, moist climates. They are also known as podzols. They occupy about 4 percent of the land surface.
- Ultisols are strongly leached, acid forest soils found primarily in humid temperate and tropical areas of the world. They have a subsurface horizon in which clays have accumulated, often with strong yellowish or reddish colors resulting from the presence of iron oxides. They are also known as lateritic soils. They occupy about 8 percent of the land surface.

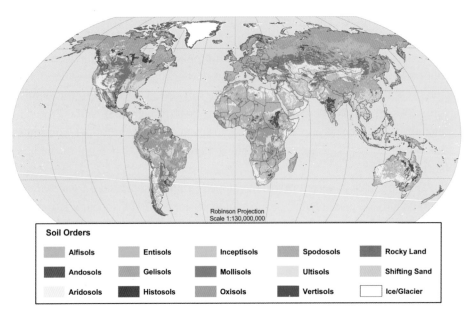

Figure 12.2 The global distribution of soil types, National Resource Conservation Service, US Department of Agriculture (source: Wikipedia, http://soils.usda.gov/use/worldsoils/mapindex/order.html).

- Vertisols are clay-rich soils, typical of savanna, that shrink and swell with changes in moisture content. During dry periods the soil volume shrinks, and deep, wide cracks form in the hard material. They lack well-developed horizons. They occupy about 2.5 percent of the land surface.

The global distribution of these soil types is shown in Figure 12.2.

12.3 Agriculture and climate

Agricultural climatology concerns the climatic resources and risks that affect crop production and the raising of livestock. Sunlight is the key control on plant photosynthesis. Photosynthesis involves the production of carbohydrates (glucose and fructose) in the green leaves using carbon dioxide and water. Light intensity, quality, and duration are important. High intensity can lead to high rates of water loss. Red light (0.7–1.4 μm wavelength) is most effective for photosynthesis. Some plants (maize, soybeans, tobacco) are short-day plants requiring only eight hours of sunlight for flowering; others (wheat, barley, oats, sugar beet) are long-day plants.

Crop growth begins at about 6 °C, but the ideal temperature is generally between 18 °C and 24 °C. Table 12.1 illustrates suitable temperature and rainfall conditions for some major crops. Most crop models make use of field-scale

Table 12.1 Suitable conditions for major crops

Crop	Temperature range (°C)	Annual rainfall (cm)
Maize	30–40	40–60
Wheat	16–25	40–150
Barley	16–25	20–100
Cotton	32–40	75–115
Sugar cane	20–30	100–150
Tobacco	15–30	40–60
Potato	15–25	50–100
Tea	20–30	150–200

weather data, but Ramirez-Villegas and Challinor show that large-scale climate data sets can be used successfully to assess climate change impacts.

Crops may be affected by freeze events, depending on their severity and the duration below a given threshold temperature. Important indicators are the average length of the frost-free season and the number of growing-degree days above some threshold. A second critical factor for crops is the water balance in the soil. This depends on the precipitation and evapotranspiration (ET) and their seasonal distribution. The Palmer Drought Severity Index (PDSI) provides a useful guide to soil moisture conditions. In areas of low and poorly distributed annual rainfall, dry farming is usually practiced, with crops such as sorghum, millet, and barley. In areas with high and regular rainfall, crops like rice, tea, coffee, and rubber are grown. Heavy rainfall is damaging because it leaches nutrients from the soil and leads to slope wash. Cool-season rainfall is beneficial as then ET is lower. Rainfall amounts in individual events need to be around 15–20 mm in order for the soil to be wetted to a depth of 10–12 cm to maintain the soil moisture. Dew has the effect of directly wetting plant leaves, which can absorb the moisture, and offsetting some of the morning evaporative loss. Fog wets the aerial parts of plants and is also associated with high relative humidity in the air.

Atmospheric humidity affects the ET rate. In dry conditions the water content of leaf tissue decreases and the plant may wilt. Moist air favors the growth of many fungi and bacteria, as well as insect parasites, which can severely affect crops. Blight diseases of potato and tea readily spread in moist atmospheric conditions.

Winds cause soil erosion and abrasion of plants, especially tall ones. Strong winds that are also hot and/or dry may cause young plants to shrivel because they increase ET losses.

For both dairy cattle and feedlot operations, heat stress when temperatures exceed 33 °C is a major concern affecting milk yield and cattle stress. At the

10 percent probability level, there are about 60 such days in Colorado, but around 80 in Texas, for example.

In the United States, climate data for applied purposes are available through the Applied Climate Information System (ACIS) developed by the six Regional Climate Centers. Information on products is given at www.rcc-acis.org/products.php

12.4 Water resources

Water is a critical and essential resource. The history of human civilization is intertwined with the history of the ways humans have learned to manipulate and use water resources. The earliest agricultural communities arose where crops could be grown with dependable rainfall and perennial rivers. Irrigation canals permitted greater crop production and longer growing seasons in dry areas, and sewer systems fostered larger population centers.

The Earth has a stock of approximately 1.4 billion km^3 of water in a wide variety of forms and locations. The vast majority ($\sim$ 97 percent) is salt water in the oceans. The world's *total freshwater reserves* are estimated at around 35 million km^3. Most of this, however, is locked up in the Antarctic and Greenland ice sheets, or in deep groundwater inaccessible to humans, leaving only 0.7 percent available for consumption. Of this amount, 87 percent is used for agriculture. Around 1.2 billion people live in areas of water scarcity, while a further 1.6 billion lack the infrastructure to take water from rivers and aquifers

Water resources are intimately linked to climate. The primary factors are precipitation, ET, soil moisture, and water storage in the ground, snow packs, and glaciers. The horizontal transfers of water in the atmosphere, and on the surface or below ground, are critical factors in the hydrological cycle that link the land and oceans. Figure 12.3 shows that 40 000 $km^3\,yr^{-1}$ are transferred from ocean to land, and a further 74 000 $km^3\,yr^{-1}$ are evaporated from the land. Ocean evaporation is 5.8 times larger than from the land, but if allowance is made for the greater fraction of the Earth that is ocean, the ratio is 2.35 times. Soil moisture is used by plants, which return the moisture to the atmosphere by transpiration. Water that does not evaporate or transpire, or seep into aquifers, runs off to form rivers and streams. Snow storage in winter in the mountains provides water for rivers in the spring and summer.

An important concept in hydrology is the *return period*, which is an estimate of the time between events like floods or droughts. It is a statistical measurement denoting the average recurrence interval over an extended period of time. For most planning purposes, a 100-year event is determined. It is determined from $(n + 1) / m$, where n is the number of years on record and m is the rank of the event being considered. A 100-year flood (or drought) has a 1 in 100 (or 1 percent)

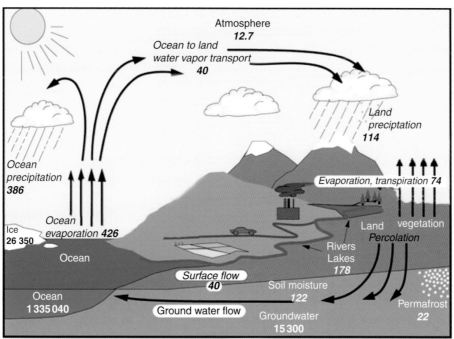

Units: Thousand cubic km for storage, and *thousand cubic km/yr* for exchanges.

Figure 12.3 The hydrologic cycle from reanalysis data, 2002–2008 (source: Trenberth and Fasullo 2012).

chance of being exceeded in any one year. It is important to stress that the 100-year interval is a long-term statistical average. Conceivably, a 500-year event could occur in the same time interval.

Among the most significant consequences of climate change are impacts on the hydrologic cycle. Changing climatic conditions will affect the supply of and demand for water resources. The IPCC reports conclude that freshwater systems are among the most vulnerable sectors. Higher temperatures are expected to intensify the hydrological cycle. In general, current climate models project that global precipitation will increase, although this is subject to significant spatial and temporal variability, with an increase in the Tropics and high latitudes and decrease in the Subtropics and mid-latitudes based on Meehl *et al.* The following are some ways in which water resources will be affected by climate change according to Gleick *et al.*:

- There will be changes in the timing and regional patterns of precipitation (very high confidence) and average precipitation will increase in higher latitudes, particularly in winter (high confidence). A growing number of studies suggest that there will be increased frequency and intensity of the heaviest precipitation events, but there is little agreement on detailed regional changes. Research suggests that flood and drought frequencies are likely to change.

- Temperature increases in mountainous areas with seasonal snowpack will lead to increases in the ratio of rain to snow and decreases in the length of the snow storage season (very high confidence).
- Higher sea levels associated with thermal expansion of the oceans and increased melting of glaciers will cause salt-water intrusion inland into rivers, deltas, and coastal aquifers (very high confidence).
- Water-quality problems will worsen where rising temperatures are the predominant climate change. Where there are changes in flow, complex positive and negative change in water quality will occur.
- The southern boundary of continuous permafrost is projected to shift north by 500 km over the next 50 years due to warming.

Because water is a fundamental element of our climate system, these changes will have important implications for the hydrologic cycle, including impacts on water availability, timing, quality, or demand. Early research on climate change and water was used to justify the implementation of various mitigation strategies. The discussion about climate change and water has since shifted to include adaptation – actions or policies that reduce vulnerability or increase resilience to inevitable climate change impacts. Adaption does not only serve to minimize our vulnerability to climate change impacts, it can also reduce our vulnerability to current climate variability and promote sustainable development and practices. As noted by Fredrick and Gleick: "The socioeconomic impacts of floods, droughts, and climate and non-climate factors affecting the supply and demand for water will depend in large part on how society adapts."

Climate change is leading to a decrease in the snow:rain ratio, an increase in ET, and accelerated melting of mountain glaciers, as well as changes in the frequency of droughts and intense rainfall events (see Section 11.2). Annual rainfall is projected to increase in the Tropics and high latitudes and to decrease in the Subtropics. Global runoff rises by about 4 percent per 1 °C of warming and, taking into account the changes in ET, global runoff is likely to increase by 7.8 percent by 2100.

The hydrologic cycle of the western United States changed significantly over the last half of the twentieth century. Barnett *et al.* provide a multivariable climate change detection and attribution study, using a high-resolution hydrologic model forced by global climate models. They focus on the changes that have already affected this primarily arid region with a large and growing population. The results show that up to 60 percent of the climate-related trends of increased winter air temperature, decreased snow pack, and increased spring and decreased summer river flow between 1950 and 1999 are human-induced.

During 1961–2003, changes in *groundwater* storage due to increasing with-drawals contributed up to 1.05 mm yr^{-1} to global sea-level rise according to one estimate. This was somewhat offset by a storage increase in dams and reservoirs of -0.39 mm yr^{-1}. The net contribution to sea level amounted to $+0.77$ mm yr^{-1} (including minor contributions from climate-driven terrestrial water storage $+0.08$, and the Aral Sea shrinkage $+0.03$ mm yr^{-1}), which explains ~ 42 percent of the observed sea-level rise, according to Pokhrel *et al.* They show that the contribution of groundwater depletion has been increasing monotonically, whereas dam construction has leveled off. In the United States groundwater depletion due to irrigated agriculture has been most severe in the Central Valley of California, western Kansas and the Texas Panhandle. The Ogallala aquifer (mainly in Nebraska, western Kansas and western Oklahoma) is being recharged at about 25 mm yr^{-1}, but there is a net overdraft of 55 mm yr^{-1}, which is clearly unsustainable. Satellite gravity measurements show that groundwater in north-western India is being depleted at an average rate of 4 cm yr^{-1} equivalent depth of water. Over the six-year study period, the groundwater depletion was 109 km^3, roughly double the capacity of India's largest reservoir.

12.5 Renewable energy

The principal climatic sources of *renewable energy* are solar and wind power. As of 2011, 20 percent of Germany's power came from renewable resources.

Solar energy is captured by solar panels that are typically a connected assembly of photovoltaic cells that convert sunlight into electricity. These are mounted on roofs and in arrays on the ground, positioned so as to maximize the energy receipt. Typical domestic output is in the range 4–10 kW and a system may save 4000–6000 kg of CO_2 emissions annually in sunny environments. Germany has 25 000 MW of installed capacity, mostly rooftop domestic installations. Commercial photovoltaic systems of ground-mounted arrays are rapidly being developed; there are over 250 "solar farms" with a power output of 10 MW or more that feed power into the electricity grids. Currently, the leading nations are Spain, Germany, and the United States. A German consortium had planned to build a solar farm in the Sahara desert of Morocco to drive steam turbines, but this scheme has for now been shelved.

Propeller-driven wind turbines generate *wind power*. Three vertically mounted propeller blades that are 20–60 m tall are mounted on towers 60–100 m high. The blades can rotate so that they face into the wind and the angle or pitch of the blades can be changed when the wind is strong enough to put undue stress on the blades. Figure 12.4 shows the average annual wind speed at a height of 80 m over the

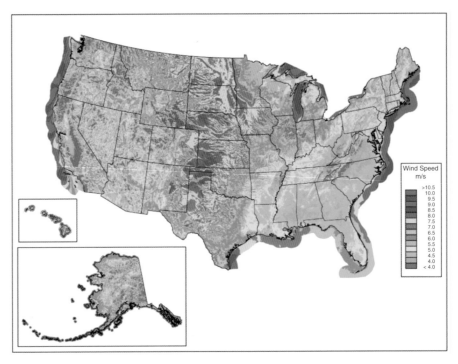

Figure 12.4 The average annual wind speed at a height of 80 m over the United States and offshore coastal zones (source: Natural Resources Energy Laboratory, US Department of Energy, www1.eere.energy.gov/wind/resource_assessment_characterization.html).

United States. There is a clear maximum over the Great Plains, at the coasts and over the Great Lakes region.

Boundary layer wind effects must be considered, because at night the blades may sweep through different near-surface wind layers. Also, there are insufficient data on wind gusts at typical turbine heights.

Installed capacity in the United States was 60 000 MW in 2012 (enough to power about 19 million homes) and accounted for almost 3.5 percent of electricity generated. China had installed 80 000 MW by 2012. Typical wind farms are 400–700 MW. Individual turbines are currently 4–8 MW capacities, but their production is generally about 35 percent of this. A measure of the wind energy available at any location is called the wind power density. It represents the mean annual power available per square meter of swept area of a turbine for a specified height. Wind farms in the United States typically have 500 W m^{-2} at 50 m height, with a mean wind speed at 50 m of 7.5 m s^{-1}. Wind power is proportional to the cube of the velocity.

Climatic assessments are essential is siting wind farms and solar farms in suitable locations. Intervals of light winds or cloudy conditions result in loss of power generation that need to be factored into the overall planning and operation of such facilities. For both these sources of renewable energy, a major factor is the need to store electricity until it is needed (at night, under cloudy skies, and in light wind conditions). Balancing electricity production from wind and solar power sources has become an important issue for planning of site installations.

12.6 Climate effects on transportation

All forms of *transportation* are affected by weather events, but also in the long term by climate. Surface transportation by road and rail is especially affected by snowfalls and freezing conditions, but also by heat waves and extreme temperatures. Extreme high temperatures can lead to rails buckling and asphalt on highways becoming soft and deforming under heavy loads.

Transportation departments use the climatological frequency of snowfalls and freezing events in planning the availability of snowplows, salt, and grit for roads ahead of each winter season. In general, disruption is greatest when snowfall or freeze events occur unexpectedly in locations where such events are rare. But even in northern latitudes large snowfall amounts (> 15 cm) may cause road closures. Rainy weather is a further transportation hazard. Traffic accidents in summer in Chicago doubled on rain days, and across the United States 27 percent of all aircraft accidents with fatalities occurred during rainy weather. In a typical year, 673 000 Americans are injured and 7000 are killed as a result of weather-related road accidents.

Warming winters have shortened the season for ice roads used by trucks to supply mining sites in northern Canada and Alaska. The opening date of ice roads in northern Alaska has been delayed by up to two months in recent years. Another effect of rising temperatures in the Arctic and Subarctic is the degradation of permafrost that is resulting in differential subsidence affecting highway embankments, railroad beds, and airport runways. The road to Umiujaq airport on the east coast of Hudson Bay, for example, subsided by > 1.2 m between 1991 and 2009 and created a wavy road surface. A significant number of Alaska's gravel airstrips are located on permafrost and hence are susceptible to the effects of thaw degradation. The Dalton Highway north of Fairbanks crosses areas of ice-rich permafrost that may thaw in future, requiring remedial reconstruction according to the US Arctic Research Commission.

Transcontinental and transoceanic aircraft operations are greatly affected by the very strong winds in jet streams in the upper troposphere. Westbound flights are typically an hour longer crossing the Atlantic or Pacific Oceans due to headwinds than eastbound flights that have tail winds. In the winter season the jet streams are stronger than in summer, at least in the Northern Hemisphere.

In spring 2010 the eruptions of an Icelandic volcano spewed massive amounts of ash into the upper atmosphere. The ash clouds, transported eastward by the jet stream, resulted in many airport closures in northern Europe during 14–20 April, and again in northwest Scotland on 3–9 May 2010, because ash can clog aircraft engines and shut them down. In 2011 there were similar airport closures in the United Kingdom on 23–25 May, following the eruption of another Icelandic volcano.

12.7 Insurance and climate/weather disasters

The frequency of billion-dollar *disasters* has greatly increased over the last two decades and the associated costs are also rising (Figure 12.5). In 2010 there were ten instances of billion-dollar weather disasters in the United States.

Currently, only about 12 percent of insurance companies take account of climate change in planning their rates for casualty and property damage. Traditionally, companies studied historical climate records to determine future risk, but with extreme events becoming more common companies may experience

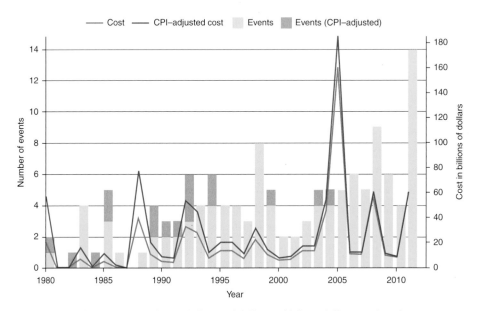

Figure 12.5 The number and cost (billions of dollars) of billion dollar weather disasters worldwide from 1980–2011. Consumer price index (CPI) adjusted values are also shown (source: NOAA).

increasing unexpected property losses, for example. The massive losses of US $41 billion resulting from the impact of Hurricane Katrina on New Orleans in 2005 led to attention to coastal hurricanes, whereas many events in 2010 were inland. In late October 2012, Hurricane Sandy tracked from the Caribbean along the east coast of the United States and severely impacted New Jersey and New York City. It was the largest hurricane on record, with a diameter of 1800 km (see the photograph which opens Chapter 11) and caused damages of US $65 billion and 253 deaths along its track.

Eighteen US states have hurricane deductibles, many of which are now changing from a fixed dollar value to a percentage of the home's value. There are also wind and hail deductibles. Flood insurance is usually a separate insurance policy, but does not cover hurricane-induced floods. An analysis of the top ten hurricane-insured losses from 1989–2005 rather surprisingly revealed that hurricanes did not have a significant impact on insurers' returns on equity.

In 2009 the National Association of Insurance Commissioners adopted a climate risk disclosure survey, which the states of California, New York, and Washington now require for large companies.

The Geneva Association, whose members represent the world's largest insurers and reinsurers, is studying the role that insurance can play in the process of adapting to the negative effects of climate change, particularly in developing countries. Renewable energy installations such as wind farms and green building construction also offer new scope for the insurance industry.

12.8 Climate forecasts and climate services

A recent development in climatology is the attention being given to *climate forecasts*. The basis of these depends very much on the time scale being considered – intraseasonal to annual, or decadal to centennial. Intraseasonal (30–90-day) variability is mostly related to intervals of zonal and blocking flow in mid-latitudes and Madden–Julian Oscillations on the Tropics. Sources of variability on this time scale are linked first to surface conditions, i.e., SSTs, land surface temperatures and soil moisture, and the extent and depth of snow cover. In high latitudes the extent and thickness of sea ice is important. A second source of variability appears to be linked to the circulation in the lower stratosphere, which has been shown to strengthen the North Atlantic Oscillation that in turn affects the winter climate over North America and Europe.

The annual interval is mainly determined by circulation modes such as ENSO, that have global impacts. It should be noted that the decay of ENSO is more predictable than its onset, which is linked to erratic and unpredictable westerly wind events in the tropical western Pacific. Another irregular factor is the occurrence of major volcanic eruptions that depress the summer temperatures for at least two years after the eruption. Decadal time scales are influenced by the solar cycle of sunspots and long-term oceanic oscillations, such as the Pacific Decadal Oscillation. Centennial trends are related to global changes initiated by rising concentrations of GHGs. Global annual temperatures are rising by 0.17 °C per decade and 2–4 times as much in the Arctic.

The Climatic Prediction Center (CPC) of the National Weather Service (NWS) of NOAA in the United States has issued seasonal outlooks since October 1995. These can be found at www.cpc.ncep.noaa.gov/products/predictions/90day

Seasonal outlooks and ENSO forecasts are also issued by the International Research Institute for Climate and Society (IRI) at Columbia University in New York. The CPC maps show the probabilities of temperature, precipitation, and SST deviations from normal for the next one- and three-month periods. These outlooks are issued from two weeks to 13 months in advance for the lower 48 states and Hawaii and other Pacific Islands. There is also a three-month drought outlook for the United States and a three-season ENSO outlook. The IRI issues three-month ENSO forecasts.

Climate forecasts are part of a growing expansion of climate services and the release of climatic data. There is an immense need for seasonal outlooks for farmers, water managers, energy utilities, ski center operators, and the construction industry. Annual outlooks are needed by these same groups as well as by relief agencies. Decadal and longer-term outlooks are needed by insurance companies, coastal protection agencies, and businesses of all kinds.

SUMMARY

Climate records are showing increases in extremes of heat and rainfall. There have been major heat waves in Europe and Russia, with significant excess mortality. Major floods have occurred in the United States and in Pakistan. Mega-droughts are reported from the American Southwest during the Medieval Climate Anomaly. There was a major climatic shift toward much drier conditions in the West African Sahel around 1970.

Soil types are closely linked to climate. A primary factor is whether P > E so that leaching occurs, or P < E so that salts are deposited on the surface. The degree of acidity is measured by the pH, which ranges from 0 to 14. Twelve different soil types are recognized by the US Department of Agriculture; the most widespread are alfisols, aridosols, entisols, and inceptisols.

Crop growth begins at about 6 °C, with an ideal temperature between 18 °C and 24 °C. Critical factors are freeze events and the soil water balance. For cattle a critical factor is heat stress at temperatures >33 °C.

Only a few percent of global water is fresh, and most of that is locked up in the major ice sheets.

The hydrologic cycle involves evapotranspiration, atmospheric transport, precipitation, and runoff; it also includes storage in the soil and snowpack. These components are particularly sensitive to climatic variability and change. Global warming is leading to a decrease in the snow:rain ratio, an increase in evapotranspiration, a shift in the timing of peak runoff, and accelerated melting of mountain glaciers. Global runoff rises by about 4 percent per 1 °C of warming. Human depletion of aquifers is contributing to sea-level rise, accounting for 42 percent of the rise since 1960.

Solar and wind power are climatic sources of renewable energy. Germany, Spain, and the United States are the leaders in photovoltaic installations (solar farms). Germany has 25 000 MW of installed capacity, mostly rooftop domestic installations. Wind power from propeller-driven turbines is being rapidly installed in China and the United States. Each country has over 60 000 MW of installed capacity.

Surface transportation is affected by climatic factors (snowfall, freezing events, and extreme high temperatures) in significant ways. Rainy weather is a factor in both highway and aircraft accidents. Warmer winters are shortening the season for ice roads in the Subarctic. Thaw degradation of permafrost is affecting the infrastructure in northern communities. Aircraft operations are greatly affected by jet streams, and ash from volcanic eruptions may affect flight operations downwind.

In 2010 the United States reported ten weather disasters with over US $1 billion losses. Currently only 12 percent of insurance companies are taking account of climate change impacts in assessing insurance rates. Policies have deductibles for hurricanes, hail, and winds.

Climate forecasts on intraseasonal to decadal scales are now being made experimentally as part of the expansion of climate services. In the United States seasonal outlooks have been made operationally since 1995.

QUESTIONS

1   Select a documented extreme climate event and summarize its characteristics, impacts, and causes.
2   What are the principal effects of climatic factors on soils?
3   What climatic factors affect crop growth in mid-latitudes?

4   Outline one of the ways in which water resources will be impacted by global warming.

5   Describe the conditions needed to maximize the utilization of wind power and solar power.

6   Evaluate the main effects of climatic factors on various forms of transportation.

7   Outline the risks for the insurance industry likely to result from climate changes projected by 2030.

8   Summarize the different time scales on which climate forecasts are being made and indicate some of the factors that are being considered.

Appendix A: Units

Système International (SI) units

The basic SI units are meter, kilogram, second (m, kg, s).

1 mm	= 0.03937 in	1 in	= 25.4 mm
1 m	= 3.2808 ft	1 ft	= 0.3048 m
1 km	= 0.6214 miles	1 mi	= 1.6090 km
1 kg	= 2.2046 lb	1 lb	= 0.4536 kg
1 m s^{-1}	$= 2.2400 \text{ mi hr}^{-1}$	1 mi hr^{-1}	$= 0.4460 \text{ m s}^{-1}$
1 m^2	$= 10.7640 \text{ ft}^2$	1 ft^2	$= 0.0929 \text{ m}^2$
1 km^2	$= 0.3861 \text{ mi}^2$	1 mi^2	$= 2.5900 \text{ km}^2$
1 °C	= 1.8 °F	1 °F	= 0.555 °C

Temperature conversions can be determined by noting that:

$$T(°F) = T(°C) \times 9/5 + 32$$

$$T(°C) = (T(°F) - 32) \times 5/9$$

and for kelvin (K),

$$273 \text{ K} = 0 °C$$

Density units $= \text{kg m}^{-3}$.

Force units: A newton (N) is equal to the force that would give a mass of one kilogram an acceleration of one meter per second per second.

Pressure units $= \text{N m}^{-2}$ ($=$ pascal, Pa); 100 Pa ($=$ hPa) $=$ 1 mb.

Mean sea level pressure $=$ 1013 mb ($=$ 1013 hPa)

Radius of the Earth $= 6.37 \times 10^6 \text{ m}$

Mean Earth–sun distance $= 1.495 \times 10^{11} \text{ m}$

Energy conversion factors: A joule is defined as the work done by a force of 1 newton acting on an object to move it a distance of 1 m in the direction the force is applied.

4.1868 joules (J) $=$ 1 calorie

$\text{J cm}^{-2} = 0.2388 \text{ cal cm}^{-2}$

watt (W) $= \text{J s}^{-1}$

$\text{W m}^{-2} = 1.433 \times 10^{-8} \text{ cal}^{-2} \text{ min}^{-1}$

$697.8 \text{ W m}^{-2} = 1 \text{ cal cm}^{-2} \text{ min}^{-1}$

Gravitational acceleration (g) $= 9.81 \text{ m s}^{-2}$

Latent heat of vaporization (at 288 K) $= 2.47 \times 10^6 \text{ J kg}^{-1}$

Latent heat of fusion (at 273 K) $= 3.33 \times 10^5 \text{ J kg}^{-1}$

Appendix B: Web links

Climate data

Diverse data sets are available at https://climatedataguide.ucar.edu

Daily data for ~9000 stations worldwide for 18 variables (Global Surface Summary of the Day) can be found at www.ncdc.noaa.gov/cgi-bin/res40.pl?page=gsod.html

Long-term mean climatological data at 30 arc second resolution (~1 km at the Equator) are available at:

- www.cru.uea.ac.uk/cru/data/hrg
- www.ncdc.noaa.gov/oa/climate/climatedata.html
- http://worldweather.wmo.int
- www.metoffice.gov.uk/climate/uk

Other data are available from:

- http://weather.unisys.com/hurricane
- http://agora.ex.nii.ac.jp/digital-typhoon/links.html.en
- http://droughtmonitor.unl.edu
- www.ncdc.noaa.gov/paleo/data.html
- www.wcc.nrcs.usda.gov/products.html
- www.greatweather.co.uk

Daily weather maps

Global and national MSLP and upper air maps, satellite imagery, radiosonde data, hydrometeorology, ocean conditions

www.weathercharts.org/indexpage2.htm

United States

www.hpc.ncep.noaa.gov/dailywxmap

Europe and North Africa 1877–1938

http://gcmd.nasa.gov/records/GCMD_NCL00001_2_4_25_26.html

Satellite data

www.star.nesdis.noaa.gov/star/smcd.php
http://earthobservatory.nasa.gov
http://earth.esa.int/dataproducts
Snow and ice data: http://nsidc.org

Earth observations

www.geoportal.org/web/guest/geo_home

Intergovernmental Panel on Climate Change

http://ipcc.ch/publications_and_data/publications_and_data_reports.shtml#.

Monsoons

www.cpc.ncep.noaa.gov/products/Global_Monsoons/Global-Monsoon.shtm

Teleconnections

www.esrl.noaa.gov/psd/data/climateindices/list
www.ncdc.noaa.gov/teleconnections/ao
www.cpc.ncep.noaa.gov/products/precip/CWlink/daily_ao_index/aao/
  month_aao_index.shtml
www.ncdc.noaa.gov/oa//climate/research/teleconnect/teleconnect.html#PDO
http://en.wikipedia.org/wiki/File:MJO_5-day_running_mean_through_1_Oct_2006.png

Crop damage

http://sciencepolicy.colorado.edu/socasp/weather1/changnon.html

Evaporation in the Amazon

www.waterandclimatechange.eu/evaporation/amazon-river-basin-evaporation-in-
  average-year

Hurricane season

http://en.wikipedia.org/wiki/2010_Atlantic_hurricane_season

Paleoclimate

www.scotese.com/miocene.htm

Glossary

Absolute vorticity   The sum of the Earth's vorticity and the relative vorticity.

Active layer   The upper layer of permafrost that thaws seasonally to a depth of between 0.5 and 3–4 m, depending on latitude and soil conditions.

Adaptation   The process of reducing the adverse effects of climate change on human and natural systems by efforts to cope with actual change, as well as by adjusting to expected change.

Adiabatic   A change in temperature that occurs without loss or gain of heat and is reversible.

Adiabatic lapse rate   The rate at which air cools on rising. Unsaturated air cools at a fixed rate of 9.8 °C km^{-1}; saturated air cools at a lower variable rate ($\sim$5–7 °C km^{-1}) as a result of the release of latent heat of condensation.

Advection   Horizontal transfer of air or water.

Aerosol   Airborne particles and/or liquid droplets and gas.

African Easterly Jet stream (AEJ)   A jet stream in the lower- middle troposphere over West Africa in summer.

Air mass   A large volume of air that is more or less homogeneous in the horizontal in terms of its temperature and water vapor content.

Albedo   The ratio of reflected to incoming solar radiation.

Algorithm   An effective method, expressed as a finite list of well-defined instructions, for calculating a function.

Altiplano   A high plateau in Bolivia averaging about 3000–3500 m.

Amelioration   An improvement.

AMSR-E   Advanced Microwave Scanning Radiometer-Earth Observing System.

Anabatic wind   A warm, shallow, daytime upslope wind due to heating contrasts.

Anemometer   An instrument for measuring wind speed.

Angular momentum   The product of a body's moment of inertia and its angular velocity.

Anomaly   The deviation of a climatic variable (such as temperature) in a given region over a specified time interval from its long-term average value for that region.

Antarctic Circumpolar Current (ACC)   A massive current system in the Southern Ocean that circles Antarctica

Anthropocene   The period since the Industrial Revolution (*c.*1800), when human activities began to affect global climate.

Anticyclone　A high-pressure system

Aphelion　The furthest position of the Earth in its orbit relative to the sun.

Apparent temperature　Temperature as perceived by a human.

Arctic Ocean　The north polar ocean that is almost completely surrounded by Eurasia and North America. It is partly covered by sea ice throughout the year and almost completely in winter.

Artificial neural network　A computational model that aims to replicate the neural structure and functioning of the human brain. It consists of an interconnected group of artificial neurons and it processes information using a connectionist approach. It is used to identify patterns in data.

Atlantic Meridional Overturning Circulation　A thermohaline circulation that involves wind-driven surface currents (the Gulf Stream) traveling polewards from the tropical Atlantic Ocean, cooling en route, and eventually sinking at high latitudes forming North Atlantic Deep Water. This dense water then flows into the ocean basins.

Atlantic multidecadal oscillation　An oscillation in Atlantic Ocean temperatures that lasts for several decades.

Atmospheric pressure　The force per unit area exerted on a surface by the weight of the air above that surface in the atmosphere.

Austral　Relating to the Southern Hemisphere.

Back trajectory analysis　The tracing backward of the airflow velocity (on six-hourly weather maps) to determine the source of the air at a given location several days earlier.

Baroclinic　An atmosphere in which the temperature and pressure surfaces intersect; an idealized frontal zone.

Barometer　An instrument for measuring air pressure.

Barotropic　An atmosphere in which the temperature and pressure surfaces are horizontal and parallel; an idealized air mass.

Beta effect　Beta is the rate of change of the Coriolis parameter (f) with latitude. This is large in low latitudes and gives rise to a deflection in the motion of tropical cyclones.

Bioclimatology　The study of the relationships between climate and the biosphere.

Biosphere　All forms of biological life on Earth; the global sum of all ecosystems.

Black body　A body that perfectly absorbs all the energy falling on it and correspondingly radiates it all away.

Blocking　The interruption of the westerly wind belt by an anticyclone that is more or less stationary.

Bora　A dynamically induced cold downslope wind.

Buoyancy   The force causing floating; the tendency of a liquid or gas to cause less dense objects to float or rise to the surface.

Carbon dioxide (CO_2)   A trace gas in the atmosphere emitted by plant and animal respiration. It is released by the burning of fossil fuels and functions as a greenhouse gas.

Carbon dioxide concentrations   The content of carbon dioxide in the atmosphere measured in parts per million by volume (ppmv).

Centripetal force   A force that makes air follow a curved path; it acts inward in a low pressure and outward in a high pressure.

Cenozoic Era   The era from 65 million years ago to the present. The term means "new life."

Chinook   *See* Foehn.

Clausius–Clapeyron relationship   The almost exponential relationship between the saturation vapor pressure and temperature.

Climate change   A change in the mean state or in the variability of climatic conditions.

Climate forecasts   Forecasts of future climatic conditions over seasons, inter-annually, decadally and longer.

Climate variability   This refers to short-term (seasonal, annual, inter-annual, several year) variations in climate.

Climate variation   The variation of global or regional climate about its mean state.

Climatic extremes   Extremes of temperature, rainfall, etc. that far exceed normal expectations of variability.

Cold front   The leading edge of a cold air mass.

Cold island   An urban area that is colder than the surrounding rural area.

Compression   Raising the pressure and reducing the volume of air.

Condensation   The phase change from water vapor to liquid.

Conduction   Heat transfer in a solid body by molecular transfer in the direction of the temperature gradient.

Congo River basin   The basin of the Congo River, located astride the Equator in west-central Africa. It is the world's second largest river basin.

Conservation of angular momentum   A decrease in the radius of rotation is balanced by an increase in velocity, and vice versa.

Conservation of energy   The first law of thermodynamics states that energy can neither be created nor destroyed (i.e., it is conserved).

Conservation of potential vorticity   The conservation of absolute vorticity in adiabatic motion.

Continental drift   The slow drift of continents on their tectonic plates over the Earth's surface.

Continentality   The magnitude of the annual temperature range, which is greatest in continental interiors.

Convection   Heat transfer in liquids or gases by turbulent eddies.

Convergence   The coming together of air at a point or in an area either by streamlines converging or slowing down.

Cordillera   North/south mountain ranges in western North and South America.

Coriolis force   An apparent force acting on air and other objects moving over the Earth's surface, deflecting them to the right (left) of their line of motion in the Northern (Southern) Hemisphere.

Coupled Model Intercomparison Project   A project to intercompare the results of coupled climate model experiments.

Cryosphere   The parts of the Earth where water is in solid form; all forms of snow, land ice, floating ice, and permafrost.

Cumulonimbus   Towering cumulus with an anvil of ice crystals extending through the troposphere.

Cyclone   A low-pressure system.

Dansgaard–Oeschger Oscillation   A climatic fluctuation averaging about 1500 years, recurring in the late glacial period.

Day length   The length of time each day from the moment the upper limb of the sun's disc appears above the horizon during sunrise to the moment when the upper limb disappears below the horizon during sunset.

Deciduous   A tree that sheds its leaves in winter.

Deforestation   Removal of forest cover by clearance or burning.

Dendrochonology   Tree-ring dating based on annual growth rings.

Dendroclimatology   The reconstruction of past climatic conditions and history from the study of tree growth rings.

Depression   A general term for a low-pressure system in the mid-latitudes or Tropics.

Derecho   An extensive, long-lived, linear windstorm that is associated with a fast-moving band of severe thunderstorms.

Desertification   A process of land degradation in desert margins.

Dew   Water droplets condensed on plant surfaces as a result of nocturnal radiative cooling.

Dew point   The temperature at which saturation occurs when air is cooled at constant pressure.

Diabatic   A change in temperature that occurs as a result of the gain or loss of heat.

Diurnal   Daily.

Disaster   A natural or man-made event of substantial extent causing significant physical damage or destruction, loss of life, or drastic change to the environment.

Divergence   The moving apart of air from a point or area either by streamline separation or speeding up.

Drought   An extended period without rainfall (meteorological drought).

Dry adiabatic lapse rate (DALR)   A constant cooling (warming) rate as unsaturated air is forced to rise (sink). It has a value of 9.8 $°C \, km^{-1}$.

Dust devil   A small-scale convective vortex that forms over strongly heated surfaces.

Dust Veil Index (DVI)   An index of stratospheric volcanic dust developed by Hubert Lamb. The index of eruptions from 1500 to 1983 is scaled to the eruption of Krakatoa (1883), which equals 1000.

Earth's orbit   The slightly elliptical path of the Earth around the sun.

Easterlies   A broad band of easterly winds in the Tropics of each hemisphere.

Easterly wave   A synoptic-scale wave disturbance in the tropical easterlies.

Eccentricity   The degree of departure from spherical in the Earth's orbit about the sun.

Eddy   A small-scale circular motion in air or water.

Electromagnetic energy   Energy that travels as wave motion at the speed of light across space without the involvement of any intervening matter. It has a spectral range from fractions of a micrometer to centimeters.

El Niño   Originally this referred to the warm current that appears in December off Peru. In the 1950s the term became applied to the suite of climatic conditions that periodically develop in the equatorial Pacific Ocean when the trade winds weaken (*see also* Southern Oscillation).

Empirical orthogonal function (EOF) analysis   The decomposition of a data set into a set of orthogonal base functions determined empirically from the data.

Ensemble (of models)   A collection of models whose results are averaged.

Evapotranspiration   The total water loss from a land surface by evaporation and transpiration from plants.

Exponential   A growth rate that increases in proportion to the current value of the function and thus increases at an accelerating rate. A negative exponential rate decreases in the same way.

Extratropical   Outside of the Tropics.

Extratropical cyclone   A low-pressure system in mid to high latitudes.

Feedback   A process in which "information" about the past or the present influences the same phenomenon in the present or future. A positive (negative) feedback amplifies (damps) the response. A negative feedback tends to be self-regulating, whereas a positive feedback enhances the original effect. Feedbacks are a major feature of the climate system.

Feedlot   A type of animal feeding operation which is used in factory farming for finishing livestock, notably beef cattle, prior to slaughter.

Ferrel cell   A thermally indirect north–south circulation cell in middle latitudes theoretically driven by the flanking direct cells equatorward and poleward.

Flood   A very high water level in rivers causing them to breach their banks and affect adjacent low-lying ground.

Flux   A flow or transfer.

Foehn   A downslope wind forced to descend dynamically that gives rise to warming and drying of the air. The term Chinook is widely used in North America for the same phenomenon.

Fog   An obscuration to visibility to less than 1 km due to suspended water droplets near the surface.

Foraminifera   Marine protozoa (benthic or planktonic) widely used in biostratigraphy.

Freezing degree day   The difference between the average daily temperature below zero and 0 °C. Daily values are summed over a month or winter season.

Friagem   A cold southerly air flow into the Amazon Basin.

Friction   The force that resists the motion of air flowing over the surface. It comprises skin friction in a molecular layer and form drag due to roughness elements in the ground (boulders, vegetation, buildings). Friction operates in the atmospheric boundary layer that is about 1 km deep.

Front   A boundary between two air masses with contrasting properties of temperature and moisture.

Frontal zone   The transition zone 50–100 km wide in the troposphere that is identified as a front on surface weather maps.

Gaussian distribution   A statistical probability distribution (also known as a normal distribution) that is bell shaped.

General Circulation Model (GCM)   A set of numerical equations specifying the global circulation and climatic conditions on horizontal and vertical grids. Initial conditions include solar radiation, atmospheric composition, land-surface and sea-surface temperatures. In coupled models the atmospheric GCM is coupled to an ocean GCM. The time step for the iteration of the numerical equation is of the order of a few minutes.

Geoengineering   The deliberate large-scale modification of climate to offset global warming.

Geological time scale   A system of chronological measurement that relates rock stratigraphy to time.

Geomorphology   The study of landforms and the processes involved in their formation.

Geopotential height   A height, adjusted for latitudinal variations of gravity, that is used to indicate the altitude of pressure levels in the atmosphere.

Geostrophic wind   The wind that is balanced by the pressure gradient force and the opposing Coriolis force, blowing parallel to straight isobars above the friction layer.

Glacial cycle   An interval comprising a glacial period (about 90 percent of the time) and an interglacial (10 percent of the time).

Global brightening   The recovery of solar radiation after the 1980s, generally as a result of decreased aerosol loading.

Global climate models   Numerical models of the global climate system that simulate all the major processes in the atmosphere and ocean.

Global dimming   An interval with reduced solar radiation as a result of aerosol loading from about 1960–1980s.

Graupel   Soft hail; refers to precipitation that forms when supercooled water droplets are collected and freeze on a falling snowflake,

Greenhouse effect   The effect in the atmosphere of warming due to the trapping of infrared radiation by greenhouse gases.

Greenhouse gas   A gas such as water vapor, carbon dioxide, or methane that absorbs infrared radiation in the atmosphere.

Grosswetter   A large-scale weather pattern.

Groundwater   Water located beneath the Earth's surface in soil pores and in the fractured rocks located beneath the Earth's surface.

Gulf Stream   A massive boundary current that flows northeastward off the eastern United States.

Haboob   A massive dust storm.

Hadley cell   A north–south vertical circulation cell in low latitudes. Air rises at the Equator and sinks in the Subtropics.

Hail   Balls or irregular lumps of ice, each of which is called a hailstone, falling to the surface. These consist mostly of water ice and measure between 5 mm and 15 cm in diameter.

Harmattan   A dry, dusty wind from the Sahara that blows toward the coast of West Africa in winter.

Haze   An obscuration to visibility by smoke or dust aerosols.

Heat flux   The flow or transfer of heat by conduction or convection.

Heat island   An urban area that is significantly warmer than its rural surroundings as a result of human activity.

Heat wave   A spell of anomalously hot weather.

Helm wind   A strong easterly wind that blows over the Northern Pennines of England in spring.

Hindcast   Data for past events are input into a climate model that is then run to allow "projections" to be compared with the actual course of the past climatic conditions.

Hoar frost   Frost that forms at night on the ground or grass.

Holocene   A geological epoch that began around 11 700 years ago and is continuing. It is part of the Quaternary period. It means "entirely recent."

Hurricane   A tropical cyclone over the Atlantic Ocean.

Hydrosphere   All forms of water on the Earth's surface and in the atmosphere.

Hydrostatic equilibrium   The balance in the atmosphere between gravity (downward) and the pressure gradient (upward).

Hygrograph   A recording device comprising human hair that charts changes in relative humidity.

Hypothermia   A condition where the body core temperature drops below that required for normal metabolism.

Hypoxia   A condition where the body is deprived of an adequate supply of oxygen.

Hypsithermal   The thermal maximum in the Holocene about 8000 to 5000 years ago

Hythergraph   A plot of monthly temperatures and precipitation amount (derived from the Greek words for water and heat).

Ice plateau   The more-or-less level surface of the upper part of ice caps and ice sheets. Most of Antarctica and Greenland are ice plateaus.

Infrared radiation   Electromagnetic radiation between about 3 and 100 μm in wavelength.

Insolation   A contraction of incoming solar radiation.

Instability line   A line of thunderstorms (e.g., ones that travel westward over West Africa in summer).

Interglacial   The warm interval between glacial periods.

Intergovernmental Panel on Climate Change (IPCC)   A large international group of climate scientists that at five-year intervals since 1990 has prepared assessment reports on climate change, its causes, and its impacts.

Inter-Ocean Convergence Zone (IOCZ)   A convergence zone in tropical southern Africa between air from the Atlantic and from the Indian Ocean.

Inter-Tropical Convergence Zone (ITCZ)   A zone of convergence between the trade wind systems of the two hemispheres.

Intertropical Discontinuity (or Front)   The boundary between monsoon air and drier continental air (e.g., over West Africa or India in summer).

Inversion   A reversal of the normal temperature decrease with height.

Isobar   A line on a map connecting points with equal pressure.

Isobaric pattern   A pattern of isobars.

Isohyet   A line on a map connecting points with equal rainfall.

Jet stream   A narrow band of very strong winds mostly located in the upper troposphere.

Katabatic wind   A cold, shallow, thermally induced, nocturnal downslope wind.

Kinetic energy   Energy due to motion.

Kuroshio   A warm ocean current flowing northeastwards off East Asia.

Lake effect   The effect of large lakes on the downwind shore leading to increased cloudiness and precipitation.

La Niña   Spanish for "the girl," it refers to the opposite cold mode of El Niño.

Last Glacial Maximum (LGM)   The last glacial phase of the Pleistocene that culminated about 21 ka.

Latent heat   Heat associated with phase changes of water from vapor to liquid, or liquid to solid, and vice versa, without a change of temperature. The processes involved are evaporation/condensation and freezing/melting.

Leeward   The lee (or downwind) side of a mountain or hill.

Lightning   A massive electrostatic discharge between positively and negatively charged regions within or between clouds, or between a cloud and the Earth's surface.

Line squall   A series of squalls along a narrow line of thunderstorms.

Little Ice Age   A cold interval particularly around the North Atlantic and in Europe from AD 1550 to 1850.

Low-level jet (LLJ)   A narrow flow of strong winds ($\sim 15\text{--}25$ m s^{-1}) at low elevations ($\sim 1\text{--}2$ km).

Madden–Julian Oscillation   An eastward-propagating oscillation in the Tropics with a time scale of 30–90 days.

Marine Isotope Stage (MIS)   A numbering system for glacial and interglacial stages based on oxygen isotopes in marine fauna.

Maritime continent   The islands of Indonesia and New Guinea.

Maunder Minimum An interval with a near absence of sunspots over AD 1645–1715.

Medieval Warm Period A warm interval over AD 950–1250.

Meridional In the direction of latitude.

Meridional cell A north–south circulation with air rising at one end and sinking at the other.

Meridional overturning circulation (MOC) The large-scale thermohaline circulation in the Atlantic Ocean.

Mesoscale A spatial scale of 10–100 km.

Mesoscale convective systems (MCS) A cluster of severe thunderstorms organized in a traveling mesoscale system that may be amorphous or linear.

Methane (CH_4) A trace gas in the atmosphere that is released by enteric fermentation and swamps. It is a potent greenhouse gas.

Microclimate Climatic differences on the scale of the plant canopy (1 cm–50 m).

Millibar The basic unit of atmospheric pressure (equal to 1 hPa [hectopascal]); one-thousandth of a bar, the approximate value of mean sea-level pressure.

Mitigation Taking actions to prevent or reduce losses of life and property from future disasters.

Mode The positive or negative phase of a pressure oscillation; the most frequent value in a frequency distribution.

Moisture flux The transfer of moisture (vertical or horizontal).

Monsoon A seasonal wind reversal usually associated with periods of heavy rainfall in summer.

Moulin A vertical shaft in an ice sheet or glacier down which water flows in the melt season.

Multiyear ice (MYI) Sea ice that survives more than two summers.

Neoglacial A cold interval in the late Holocene when glaciers readvanced, most prominently in the Little Ice Age.

Net radiation The balance of incoming and outgoing all-wavelength radiation.

North Atlantic Oscillation (NAO) A pressure oscillation, which in its positive phase has a deep Icelandic low and a strong Azores high.

Northern Annular Mode (NAM) A pressure oscillation, which in its positive mode has low pressure in high latitudes and high pressure in mid-latitudes (and vice versa for negative mode). Also call the Arctic Oscillation (AO).

North Pacific Oscillation (NPO) A pressure oscillation, which in its positive phase has a deep Aleutian low and a strong North Pacific high.

Obliquity of the ecliptic The axial tilt of the Earth, currently 23.5°.

Occluded front A frontal boundary where the cold front has caught up with the warm front, lifting the warm sector off the ground.

Ocean acidification   A decrease in the pH of ocean water as a result of carbon uptake, with a result of the oceans becoming more acidic.

Oceanic gyre   A large horizontal circulation in the ocean.

Optical depth   A measure of transparency of the atmosphere.

Orographic uplift   The forced uplift of air by a mountain.

Oscillation   A pressure seesaw between two distant locations or regions.

Ozone   The tri-atomic form of oxygen (O_3) that forms a layer in the lower stratosphere, absorbing ultraviolet radiation. It is highly corrosive and poisonous.

Pacific Decadal Oscillation (PDO)   A multidecadal oscillation in sea-surface temperature between the eastern and western North Pacific Ocean.

Pacific–North America (PNA) pattern   A teleconnection pattern involving, in the positive phase, above-average heights in the vicinity of Hawaii and over the intermountain region of North America, and below-average heights located south of the Aleutian Islands and over the southeastern United States.

Paleoclimatology   The study of past climatic conditions from proxy records and climate model reconstructions.

Perennial   Lasting for an indefinitely long time.

Periglacial   An environment that is adjacent to ice bodies; more generally applied to cold climate processes.

Perihelion   The closest position of the Earth in its orbit relative to the sun.

Permafrost   Permanently frozen ground; it may or may not contain ice.

Photosynthesis   The process in plants that uses light together with carbon dioxide and water to form organic compounds and sugars.

Photovoltaic cell   A panel that converts sunlight into electricity. The word combines photon (for light) with voltage (for electricity).

Planetary wave   A slow-moving wave in the atmospheric circulation with a wavelength > 3000 km. Also known as a Rossby wave.

Plate tectonics   A scientific theory that describes the large-scale motions of the Earth's lithosphere.

Polar amplification   The amplification of temperature trends in the polar regions as a result of various feedback processes.

Polar deserts   Barren areas in polar regions where very little precipitation falls.

Polar front   The frontal zone separating polar and tropical air masses.

Polar front jet stream   A jet stream associated with the polar frontal zone.

Polar low   A subsynoptic-scale low-pressure system that forms over high-latitude oceans, especially in winter.

Polynya   A large, irregular opening in sea ice.

Potential energy   Energy due to position, typically as a result of gravity.

Potential evaporation   The maximum possible evaporation from a given surface that is fully supplied with water.

Precession of the equinox   The shift in timing of the vernal equinox from perihelion to aphelion over a 26 000-year cycle (effectively 21 000 years due to the combined effect of precession and orbital eccentricity).

Precipitable water   The total vapor content in an atmospheric column.

Precipitable water content   The depth of water (mm or cm) in an atmospheric column if all the water in that column were precipitated as rain.

Precipitation   All forms of liquid and solid water (hydrometeors) falling to the surface.

Proxy record   A biological, chemical, or physical characteristic that indicates past environments or climate.

Pyranometer   A thermopile covered by a glass hemispheric dome that is used to measure solar radiation.

Quasi-stationary front   A front that is moving at a speed of < 2.5 m s^{-1}.

Quaternary Period   The latest (fourth) period of geologic time beginning about 2.6 million years ago and continuing to the present.

Radar   Radio Detection and Ranging. An active microwave system emits an electromagnetic pulse that is returned by whatever surfaces it encounters. Weather radars determine cloud and precipitation structure. Satellite radars measure surface properties.

Radiation   The transfer of energy as electromagnetic waves.

Radiative cooling   Cooling by outgoing infrared radiation.

Radiative forcing   The difference between the energy received by the Earth and that radiated back to space. It can be calculated for individual greenhouse gases and aerosols, etc.

Rainshadow   An area of low precipitation in the lee of a mountain range that reduces moisture transport.

Reanalysis   The production of multiyear, global, state-of-the-art, gridded representations of atmospheric states, generated by a constant numerical model and constant data assimilation system.

Renewable energy   Energy that comes from resources which are continually replenished, such as solar radiation, wind, tides, waves, and geothermal heat.

Return period   An estimate of the likelihood of an event, such as a flood or extreme rainfall, to occur. It is a statistical measurement typically based on historic data denoting the average recurrence interval over an extended period of time.

Rossby wave   *See* Planetary wave.

Rotor   A small-scale overturning circulation in the direction of the airflow.

Salinity   The saltiness or dissolved salt content (sodium chloride, magnesium and calcium sulfates, and bicarbonates) of seawater.

Saturated adiabatic lapse rate (SALR)   A variable cooling rate for saturated air that is forced to rise. It ranges typically from about 5 to 8 °C km^{-1} as a result of the variable release of latent heat.

Scatterometer   A device designed to determine the normalized radar cross-section of the surface using pulses emitted by the sensor.

Scenario   Socio-economic projection of future greenhouse gas concentrations based on population growth, economic activity, governance structure, social values, and level of technology.

Scirocco   A hot, dry wind from the Sahara over North Africa.

Sea ice   A floating layer of frozen sea water.

Sea-level rise   The rise of global sea level due to ocean thermal expansion and the melting of glaciers, ice caps, and ice sheets.

Sea-surface temperature (SST)   The temperature of the surface layer of the ocean.

Self-organizing map (SOM)   A statistical technique based on artificial neural networks designed to cluster weather maps into types.

Sensible heat   Heat that is exchanged between a surface and the air (or vice versa) that leads to a change in temperature.

Siberian high   A shallow but intense anticyclone over eastern Siberia in winter.

Sine   A trigonometric function of an angle. It is defined in the context of a right-angled triangle as the ratio of the length of the side opposite that angle to the length of the hypotenuse.

Snowball Earth   An interval in the Precambrian when the Earth was mostly covered by land ice or sea ice.

Snow cover   A layer of snow on the ground, or the amount of an area that is covered by snow, usually given as a percentage of the total area.

Soils   Layers above the bedrock that are primarily composed of minerals, mixed with organic matter, which differ from their parent materials in their texture, structure, consistency, color, chemical, biological, and other characteristics.

Solar energy   Light and heat from the sun that is harnessed by humans using a range of technologies including solar heating of water and photovoltaics to generate electricity.

Solar radiation   Electromagnetic energy from the sun per unit area at the top of the atmosphere or at the ground surface during a given time interval. It comprises ultraviolet light, visible light, and infrared radiation.

South Atlantic Convergence Zone (SACZ)   A convergence zone extending from the Brazilian heat source southeastward into the South Atlantic ocean associated with cloud cover and precipitation.

South Pacific Convergence Zone (SPCZ)   A convergence zone extending from New Guinea southeastward into the Southwest Pacific; the northwestern part has tropical and the southeastern part frontal characteristics.

Southern Annular Mode (SAM)   A pressure oscillation, which in its positive phase has low pressure in high southern latitudes, and high pressure in mid-latitudes (and vice versa in its negative mode). Also called the Antarctic Oscillation (AAO).

Southern Ocean   The southernmost waters of the world ocean, generally taken to be south of 60° S latitude and encircling Antarctica.

Southern Oscillation (SO)   An east–west oscillation in the atmosphere over the equatorial Pacific Ocean. It is the atmospheric part of the coupled El Niño–Southern Oscillation (ENSO) phenomenon (*see also* El Niño).

Spatial resolution   The scale at which horizontal space is represented in a model.

Spatial Synoptic Classification   A classification based on air mass characteristics on a local scale.

Spectrometer   An instrument used to measure the properties of radiation over a specific portion of the electromagnetic spectrum.

Spectroradiometer   An instrument used to measure the spectral power distributions of solar radiation.

Stable air   Air is stable when, after a vertical displacement up or down, it returns to its original level.

Standard atmosphere   A series of models that give values for pressure, density, and temperature over a range of altitudes.

Standing wave   A wave in the airflow set up in the lee of a mountain that remains in place.

Steppe   A Russian term for grassland (synonymous with prairie).

Steric effect   The effect of temperature and salinity on changing the ocean volume.

Stratiform clouds   Layer clouds such as stratus and stratocumulus.

Stratosphere   Layer in the atmosphere above the tropopause where temperatures increase with height due to the absorption of solar radiation by ozone. It extends between about 10 and 50 km altitude.

Streamline   A line of instantaneous motion of the air.

Sublimation   The phase change of snow or ice directly to vapor.

Subtropical jet stream   A jet stream associated with a strong horizontal temperature gradient in the upper troposphere.

Subsidence   Sinking motion in the atmosphere.

Surface roughness   A measure of the roughness of a surface used in boundary layer studies that is about one-tenth of the canopy height.

Synoptic   Instantaneous state of the atmosphere over a wide area.

Synoptic scale   Horizontal dimensions of an extratropical cyclone.

Taiga   A Russian term for boreal forest.

Teleconnection   A distant correlation in atmospheric pressure patterns.

Thermal low   A shallow low-pressure system formed over an extensive heated surface.

Thermal wind   A component of the wind (not an actual wind) related to the vertical wind shear caused by a horizontal temperature gradient.

Thermocline   A sharp vertical temperature gradient in a water mass.

Thermohaline circulation   A circulation in the ocean driven by contrasts in temperature and salinity (and therefore density differences).

Thermometer   An instrument for measuring temperature.

Topoclimate   The climate caused by topography on a scale of hundreds of meters.

Tornado Alley   A media term for a loosely defined area of high tornado frequency in the central United States; as well as a core area over Texas, Oklahoma, and Kansas, it may include the Upper Midwest, the Ohio Valley, the Tennessee Valley, and the lower Mississippi valley.

Total freshwater reserves   The largest quantity is frozen in the Antarctic and Greenland ice sheets. The largest liquid water reserve is in the Great Lakes.

Trade wind   An easterly wind in the Tropics; there is a belt of northeasterly (southeasterly) trade winds in the Northern (Southern) Hemisphere.

Trajectory   The actual path of an air parcel.

Transportation   The movement of people, animals, and goods from one location to another by road, rail, water, air, or pipeline.

Tropical cyclone   An intense low-pressure system in low latitudes, characterized by an eye in the center.

Tropopause   The upper limit of the troposphere, which is marked by a temperature increase with height. It is at about 8 km nears the Poles and 20 km in the Tropics.

Troposphere   The lowest layer of the atmosphere where almost all weather phenomena occur. Tropos is Greek for "mixing."

Tropospheric river   A narrow band of concentrated moisture transport in the atmosphere typically several thousand kilometers long and only a few hundred kilometers wide.

Tundra A Lapp word meaning treeless, applied to areas north of the treeline in the Arctic or above the treeline in the Alpine.

Turbulence Small-scale eddy motion in the atmosphere.

Typhoon A tropical cyclone in the western Pacific Ocean.

Ultraviolet radiation The shortest wavelengths of solar radiation reaching the Earth's surface ($< 0.4 \mu m$).

Unstable air Air is unstable when after a small upward displacement it continues to rise.

Upwelling A rising ocean water mass.

Vaporization The change of water into water vapor.

Vapor pressure The partial pressure exerted by water vapor as a gas.

Vasoconstriction The shrinking of blood vessels when cold to lessen heat loss.

Vasodilation The expansion of blood vessels when warm to increase heat loss.

Vector A physical quantity that has magnitude and direction, such as wind.

Vertical shear A change in wind direction (or speed) with altitude.

Visibility A measure of the distance at which an object (by day) or a light (by night) can be clearly discerned.

Vortex A region within the atmosphere or ocean where there is spinning motion about an imaginary axis.

Vorticity The curl of air motion; the vertical relative vorticity is that about a vertical axis.

Walker circulation A large east–west circulation cell in the Tropics.

Warm front The leading edge of a warm air mass.

Water balance The balance between precipitation, evaporation, runoff, and storage in the soil or a snowpack.

Water resources Sources of water that are useful or potentially useful for humans and animals.

Wavelength The distance between two wave crests or troughs in a successive wave pattern.

Weather satellite A satellite that is primarily used to monitor the Earth's weather.

Westerly wind burst (WWB) A brief interval (days) of enhanced westerly winds over the equatorial Pacific Ocean.

Wind chill index A temperature scale that has been adjusted to take account of the cooling effect of the wind on the skin.

Wind power The conversion of wind energy into a useful form of energy, for example, by using wind turbines to generate electrical power.

Wind shear   A change in wind velocity in either the vertical or horizontal direction.

Wind velocity   A vector quantity for the speed and direction of the wind (often it is used interchangeably with wind speed).

Windward   The upwind side of a mountain or hill.

Younger Dryas   A brief period of cold climate from about 12 900 to 11 500 BP that interrupted the postglacial warming and is attributed to the shutdown of the Atlantic conveyor belt by cold, fresh surface water.

Zonal   In the direction of longitude.

Bibliography

Allen, R. J. 2012. Recent Northern Hemisphere tropical expansion primarily driven by black carbon and tropospheric ozone. *Nature*, DOI 10.1038/nature11097.

Arguez, A. and Vose, R. S. 2011. The definition of the standard WMO climate normal. *Bull. Amer. Met. Soc.*, 92(6): 699–704.

Armstrong R., Brodzik, M. J., and Savoie, M. H., 2003. *Multi-sensor Approach to Mapping Snow Cover Using Data from NASA's EOS Aqua and Terra Spacecraft (AMSR-E and MODIS)*. Boulder, CO: National Snow and Ice Data Center (NSIDC), University of Colorado.

Baldocchi, D., Falge, E., Lianhong, G., *et al.* 2011. FLUXNET: a new tool to study the temporal and spatial variability of ecosystem-scale carbon dioxide, water vapor and energy flux densities. *Bull. Amer. Met. Soc.*, 82: 4216–34.

Barnett, T. P., Pierce, D. W., Hidalgo, H. G., *et al.* 2008. Human-induced changes in the hydrology of the western United States. *Science*, 319: 1080–83.

Barrett, E. C. 1974. *Climatology From Satellites*. London: Routledge.

Barry, R. G. 1967. Models in meteorology and climatology. In Chorley, R. J. and Haggett, P., *Models in Geography*. London: Methuen.

Barry, R. G. 2008. *Mountain Weather and Climate*. 3rd edn. Cambridge: Cambridge University Press, 512 pp.

Barry, R. G. 2012. A brief history of the terms climate and climatology. *Int. J. Climatol.*, 32: DOI: 10.1002/joc.3504

Barry, R. G. and Carleton, A. M. 2001. *Synoptic and Dynamic Climatology*. London: Routledge. 620 pp.

Barry, R. G. and Chorley, R. J. 2010. *Atmosphere, Weather and Climate*. 9th edn. London: Routledge. 516 pp.

Barry, R. G. and Gan, T. Y. 2011. *The Global Cryosphere: Past, Present and Future*. Cambridge: Cambridge University Press. 477 pp.

Barry, R. G. and Perry, A. H. 1973. *Synoptic Climatology: Methods and Applications*. London: Methuen. 555 pp.

Barry, R. G. and Perry, A. H. 2001. Synoptic climatology and its applications. In Barry, R. G. and Carleton, A. M., *Synoptic and Dynamic Climatology*. London: Routledge, pp. 547–603.

Bartholomew, J. G. 1897–1899. *Atlas of Meteorology*. London: J. Bartholomew & Co. 34 sheets.

Bartzokas, A. and Metaxas, D. A. 1996. Northern Hemisphere gross circulation types: Climate change and temperature distribution. *Met. Zeitschr.*, 5: 99–109.

Bates, B. C., Kundzewicz. Z. W., Wu, S. and Palutikof, J. P. (eds.) 2008. *Climate Change and Water*. Geneva: IPCC Secretariat, 210 pp.

Beck, C., Jacobeit, J., and Jones, P. D. 2007. Frequency and within-type variations of large scale circulation types and their effects on low-frequency climate variability in Central Europe since 1780. *Int. J. Climatol.*, 27: 473–91.

Berry, G., Jakob, C., and Reeder, M. 2011. Recent global trends in atmospheric fronts. *Geophys. Res. Lett.*, 38: L21812.

Blundon, J. and Arndt, D. S. (eds.) 2012. State of the climate in 2012. Special supplement to *Bull. Amer. Met. Soc.*, 93(7): 1–263.

Bonan, G. B. 1995. Sensitivity of a GCM simulation to inclusion of inland water surfaces. *J. Climate*, 8: 2691–704.

Bradley, R. S. 1999. *Paleoclimatology: Reconstructing Climates of the Quaternary*. New York: Academic Press.

Brown, W. 1842. Mean pressure of the atmosphere in different latitudes. *London, Edinburgh and Dublin Phil. Mag. and J. Science*, 20: 469.

Buchan, A. 1867. The mean pressure of the atmosphere and the prevailing winds over the globe for the months and for the year. Part 1. January, July. Year. *Proc. Roy. Soc. Edin.*, 6: 303–7.

Buchan, A. 1869. The mean pressure of the atmosphere and the prevailing winds over the globe for the months and for the year. Part 2. *Proc. Roy. Soc. Edin.*, 25: 575–637.

Budyko, M. I. 1962. The heat balance of the surface of the earth. *Soviet Geography* 3(5): 3–16.

Budyko, M. I. 1974. *Climate and Life*. (Transl. D. H. Miller). New York: Academic Press, 346 pp.

Bulygina, O. N., Razuvaev, V. N., and Korshunova, N. N. 2009. Changes in snow cover over Northern Eurasia in the last few decades. *Environ. Res. Lett.*, 4: 045026.

Burnette, D. J. and Stahle, D. W. 2013. Historical perspective on the dust bowl drought in the central United States. *Clim. Change*, 116: 479–94.

Carleton, A. M. 1991. *Satellite Remote Sensing in Climatology*. London: Belhaven. 291 pp.

Cassano, J. J., Uotila, J., and Lynch, A. 2006. Changes in synoptic weather patterns in the polar regions in the twenty-first century. Part 1: Arctic. *Int. J. Climatol.*, 26: 1027–49.

Cassano, E. N., Cassano, J. J., and Nolan, M. 2011. Synoptic weather pattern controls on temperature in Alaska. *J. Geophys. Res.*, 116: D11108.

Cassou, C., Terray, L., Hurrell, J. W., and Deser, C. 2004. North Atlantic winter climate regimes: spatial asymmetry, stationarity with time, and oceanic forcing. *J. Climate*, 17: 1055–68.

Catto, J. L., Jakob, C., Berry., G., and Nicholls, N. 2012. Relating global precipitation to atmospheric fronts. *Geophys. Res. Lett.*, 39: L10805. DOI:10.1029/2012GL051736.

Cecil, D. J. and Blankenship, C. B. 2012. Toward a global climatology of severe hailstorms as estimated by satellite passive microwave imagers. *J. Climate*, 25: 687–703.

Chagnon, S. 1996. Effects of summer precipitation on urban transportation. *Clim. Change*, 32: 481–94.

Chandler, T. J. 1965. *The Climate of London*. London: Hutchinson. 292 pp.

Coffin, J. H. 1853. *On the Winds of the Northern Hemisphere*. Washington, DC: Smithsonian Institute.

Coffin, J. H. 1876. *The Winds of the Globe or the Laws of the Atmosphere's Circulation over the Surface of the Earth*. Washington, DC: Smithsonian Institute. 756 pp.

Coleman, J. S.M. and Rogers, J.C, 2007. A synoptic climatology of the central United States and associations with Pacific teleconnection pattern frequency. *J. Climate*, 20: 3485–97.

Cook, E. R., Seager, R., Heim, R. R., Vose, R. S., Herweijer, C., and Woodhouse, C. 2009. Megadroughts in North America: Placing IPCC projections of hydroclimatic change in a long-term paleoclimate context. *J. Quatern. Sci.*, DOI: 10.1002/jqs.1303.

Coumou, D. and Rahmstorf, S. 2012. A decade of weather extremes, *Nature Clim. Change*, 2(7): 491–96.

Crane, R. G. and Barry, R. G., 1988. Comparison of the MSL synoptic pressure patterns of the Arctic as observed and simulated by the GISS General Circulation Model. *Met. Atmos. Phys.*, 39: 169–83.

Crowe, P. R. 1971. *Concepts of Climatology*. London: Longmans. 589 pp.

Dai, A. 2010. Drought under global warming: A review. *Wiley Interdisc. Rev. Clim. Change*, 2: 45–65.

Daly, C., Widrlechner, M. P., Halbleib, M. D., Smith, J. I., and Gibson, W. P. 2012. Development of a new USDA plant hardiness zone map for the United States. *J. Appl. Met. Clim.*, 52: 242–64.

DeConto, R. M., Galeotti, S., Pagani, M., *et al.* 2012. Past extreme warming events linked to massive carbon release from thawing permafrost. *Nature*, 484: 87–91.

Diaz, H. F. and Eisacheid, J. K. 2007. Disappearing "alpine tundra" Köppen climatic type in the western United States. *Geophys. Res. Lett.*, 34: L18707.

Diaz, H. F. and Kiladis, G. N. 1992. Atmospheric teleconnections associated with the extreme phases of the Southern Oscillation. In H. F. Diaz and V. Markgraf (eds.), *El Niño: Historical and Paleoclimatic Aspects of the Southern Oscillation*. Cambridge: Cambridge University Press, pp. 7–28.

Diaz, H. F., Trigo, R., Hughes, M., Mann, M., Xoplaki, E., and Barriopedro, D. 2011. Spatial and temporal characteristics of climate in medieval times revisited. *Bull. Amer. Met. Soc.*, 92: 1487–500.

Dixon, P. G., Brommer, D. M., Hedquist, B. C., *et al.* 2005. Heat mortality versus cold mortality. *Bull. Amer. Met. Soc.*, 86: 937–43.

Dzerdeevski, B. L. 1963. Fluctuations of general circulation of the atmosphere and climate in the twentieth century. In: *Changes of Climate: Arid zone Research 20*: Paris: UNESCO, pp. 285–95.

Eddy, J. A. 1976. The Maunder minimum. *Science*, 192(4245): 1189–1202.

Emanuel, K. 2003. Tropical cyclones. *Ann. Rev. Earth Planet. Sci.*, 31: 75–104.

Fasullo, J. T. and Trenberth, K. E. 2008. The annual cycle of the energy budget. Part II: Meridional structures and poleward transports. *J. Climate*, 21: 2313–25.

Fischer, E. M., Luterbacher, J., Zorita, E., Tett, S. F. B., Casty, C., and Wanner, H. 2007. European climate response to tropical volcanic eruptions over the last half millennium. *Geophys. Res. Lett.*, 34: L05707.

Food and Agricultural Organization. 2010. *Global Forest Resource Assessment 2010*. Rome: FAO. 378 pp.

Frederick, K. D. and Gleick, P. H. 1999. *Water and Global Climate Change: Potential Impacts on U. S. Water Resources*. Washington, DC: Pew Center on Global Climate Change.

Galameau, T. J., Jr., Bosart, L. F., Aiyyer, A. R., and Atallah, E. 2004. Global climatology of 1000–500 hPa thickness highs and lows. Frederick Sanders Symposium, American Meteorological Society, January 12, 2004, Seattle, WA.

Garreaud, R., Vuille, M., and Clements, A. C. 2003. The climate of the Altiplano: Observed current conditions and mechanisms of past changes. *Palaeogeog., Palaeoclimatol., Palaeoecol.*, 194: 5–22.

Geerts, B. 2002. Empirical Estimation of the annual range of monthly-mean temperatures. *Theoret. Appl. Climatol.*, 73: 107–32.

Geiger, R., Aron, R., and Todhunter, P. 2003. *Climate Near the Ground*. 6th edn. Lanham, MD: Rowman and Littlefield. 584 pp.

Gimeno, L., Trigo, R. M., and Stohl, A. 2010. On the origin of continental precipitation. *Geophys. Res. Lett.*, 37: L13804, DOI:10.1029/ 2010GL043712.

Gimeno, L., Drumond, A., Nieto, R., *et al.* 2011. A close look at oceanic sources of continental precipitation. *Eos. Trans. Amer. Geophys. Union*, 92(23): 193–4.

Gimeno, L., Stohl, A., Trigo, R. M., *et al.* 2012. Oceanic and terrestrial sources of continental precipitation. *Rev. Geophys.*, DOI:10.1029/2012RG000389.

Girs, A. A. 1966. Intra-periodical transformations of the atmosphere and their causes. In Girs, A. A. and Dydina, L. A. (eds.), *Contributions to Long-Range Weather Forecasting in the Arctic*. Jerusalem: Israel Program of Scientific Translations, pp. 13–45.

Gleick, P. H. 2000. *Water: The Potential Consequences of Climate Variability and Change for the Water Resources of the United States*. Oakland, CA: Pacific Institute for Studies in Development, Environment, and Security. 151 pp.

Gleick, P. H. (ed.) 2009. *The World's Water, Vol. 7, 2008–2009*, Washington, DC: Island Press.

Godske, C L. 1966. Methods of statistics and some applications to climatology. In: *Statistical Analysis and Prognosis in Meteorology*. Geneva: World Meteorological Organization, pp. 9–86.

Goklany, I. M. 2007. *Death and Death Rates Due to Extreme Weather Events*. London: International Policy Press. 15 pp.

Griffin, D., Woodhouse, C., Meko, D., *et al.* 2013. North American monsoon precipitation reconstructed from tree-ring latewood. *Geophys. Res. Lett.*, 40(5): DOI: 10.1002/ grl.50184.

Grumm, R. H. 2011. The Central European and Russian heat event of July–August 2010. *Bull. Amer. Met. Soc.*, 92: 1286–96.

Grundstein, A. and Dowd, J. 2011. Trends in extreme apparent temperatures over the United States, 1949–2010. *J. Appl. Met. Climatol.*, 50(8): 1650–53.

Hanson, H. P. 1991. Marine stratocumulus climatologies. *Int. J. Climatol.*, 11: 147–64.

Hartmann, D. L. 1994. *Global Physical Climatology*. San Diego, CA: Academic Press. 411 pp.

Hatzianastassiou, N., Matsoukas, C., Fotiadi, A., *et al.* 2005. Global distribution of Earth's surface shortwave radiation budget. *Atmos. Chem. Phys. Discuss.*, 5: 4545–97.

Hays, J. D., Imbrie, J., and Shackleton, N. J. 1976. Variations in the Earth's orbit: Pacemaker of the Ice Ages. *Science*, 194(4270): 1121–32.

Hewitson, B. C. and Crane, R. G. 2002. Self-organizing maps: applications to synoptic climatology. *Clim. Res.*, 22: 13–28.

Höppe, P. 1997. Aspects of human biometeorology in past, present and future. *Int. J. Biomet.*, 40: 19–23.

Howard, L. 1818–1820. *The Climate of London, Deduced from Meteorological Observations, Made at Different Places in the Neighbourhood of the Metropolis*. 2 vols., London: W. Phillips.

Hsu, P., Li, T., and Wang, B. 2011. Trends in global monsoon area and precipitation over the past 30 years. *Geophys. Res. Lett.*, 38: L08701.

Hubbard, K. G. 2007. Agricultural climatology. *J. Service Climatol.*, 1: 1–9.

Huffman, G. J., Adler, R. F., Arkin, P., *et al.* 1997. The Global Precipitation Climatology Project (GPCP) combined precipitation dataset. *Bull. Amer. Met. Soc.*, 76: 5–20.

Huth, R., Beck, C., Philipp, A., *et al.* 2008. Classifications of atmospheric circulation patterns: Recent advances and applications. *Ann. N. Y. Acad. Sci.*, 1146: 105–52.

James, P. M. 2007. An objective classification method for Hess and Brezowsky Grosswetterlagen over Europe. *Theoret. Appl. Clim.*, 88: 17–42.

Johnson, N. C., Feldstein, S. B., and Tremblay, B. 2008. The continuum of Northern Hemisphere teleconnection patterns and a description of the NAO shift with the use of self-organizing maps. *J. Climate*, 21: 6354–71.

Jones, J. M., Fogt, L., Widmann, M., Marshall, G., Jones, P., and Visbeck, M. 2009. Historical SAM variability. Part I: Century-length seasonal reconstructions. *J. Climate*, 22: 5319–45.

Jones, P. D., New, M., Parker, D., Martin, S., and Rigor, I. 1999. Surface air temperature and its changes over the past 150 years. *Rev. Geophys.*, 37: 173–99.

Jones, P. D., Lister, D. H., Osborn, T., Harpham, C., Salmon, M., and Morice, C. 2012a. Hemispheric and large-scale land-surface air temperature variations: An extensive revision and an update to 2010. *Rev. Geophys.*, 117: D05127.

Jones, P. D., Harpham, C., and Briffa, K. R. 2012b. Lamb weather types derived from reanalysis products. *Int. J. Climatol.*, 33: 1129–39.

Kalkstein, L. S and Corrigan, P. 1986. A synoptic climatological approach for geographical analysis: Assessment of sulfur dioxide concentrations. *Ann. Assoc. Amer. Geogr.*, 76: 381–95.

Kendrew, W. G. 1961. *The Climates of the Continents*. 5th edn. Oxford: Clarendon Press.

Kidder, S. Q and Vonder Haar, T. H. 1995. *Satellite Meteorology: An Introduction*. San Diego, CA: Academic Press. 466 pp.

Kirchhofer, W., 1973. Classification of 500 mb patterns. *Arbeit der Schweizer Meteorol. Zentral.*, 43: n.p.

Klyashtorin, L. B. 2001. *Climate Change and Long-term Fluctuations of Commercial Catches: The Possibility of Forecasting*. Rome: FAO; Fisheries Technical Paper No. 46.

Köppen, W. 1923. *Die Klimate der Erde: Gundriss der Klimakunde*. Berlin: Walter de Gruyter Co.

Köppen, W. and Geiger, E. 1930. *Handbuch der Klimatologie*, 5 vols. Berlin: Gebruder Borntraeger.

Kozuchowsky, K. and Marciniak, K. 1988. Variability of mean monthly temperature and semi-annual precipitation totals in Europe in relation to hemispheric circulation parameters. *J. Climatol.*, 8: 191–9.

Kraus, H. and Alkhalal, A. 1995. Characteristic surface energy budgets for different climate types. *Int. J. Climatol.*, 15: 275–84.

Kysely, J. and Domonkos, P. 2006. Recent increase in persistence of atmospheric circulation over Europe: comparison with long-term variations since 1881. *Int. J. Climatol.*, 26: 461–83.

Lamb, H. H. 1972. *British Isles Weather Types and a Register of Daily Sequence of Circulation Patterns, 1861–1971*. London: HMSO. 85 pp.

Lamb, H. H. 1990. *Climate: Past, Present and Future. Vol.2*. London: Methuen.

Landsberg, H. E. (ed.) 1969–1995. *World Survey of Climatology*. 16 vols. Amsterdam: Elsevier.

Lau, W. K.M. and Kim, K.-M. 2012. The 2010 Pakistan flood and Russian heat wave: Teleconnection of hydrometeorological extremes. *J. Hydromet.*, 13: 392–403.

Lavers, D. A., Allan, R., Wood, E., *et al.* 2011. Winter floods in Britain are connected to atmospheric rivers. *Geophys. Res. Lett.*, 38: L23803, DOI:10.1029/2011GL049783.

Leigh, E. G., Jr. 1975. Structure and climate in tropical rain forest. *Ann. Rev. Ecol. Systematics*, 6: 67–86

Linkin, M. E. and Nigam, S. 2008, The North Pacific Oscillation–West Pacific teleconnection pattern: Mature-phase structure and winter impacts. *J. Climate*, 21: 1979–97.

Lockwood, M., Harrison, R. G., Woolings, T., and Solanki, S.K. 2010. Are cold winters in Europe associated with low solar activity? *Env. Res. Lett.*, 5: n.p.

Lowry, W. P. 1967. *Weather and life: An Introduction to Biometeorology*. New York: Academic Press. 305 pp.

Mann, M. E., Bradley, R. S., and Hughes, M. K. 1999. Northern Hemisphere temperatures during the past millennium: Inferences, uncertainties, and limitations. *Geophys. Res. Lett.*, 26: 759–62.

Mann, M. E., Zhang, Z., Rutherford, S., *et al*. 2009. Global signatures and dynamical origins of the Little Ice Age and Medieval Climate Anomaly. *Science* 326(5957): 1256–60.

Mantua, N. 2002. The Pacific decadal oscillation. *J. Phys. Oceanog.*, 58: 35–44.

Marengo, J. A., Liebmann, B., Grimm, M. A., *et al*. 2010. Recent developments on the South American monsoon system. *Int. J. Climatol.*, 30: DOI: 10.1002/joc.2254.

McGuffie, K. and Henderson-Sellers, A. 2005. *A Climate Modelling Primer*. 3rd edn. Chichester: Wiley.

Melles, M., Brigham-Grette, J., Minyuk, P. S., *et al*. 2012. 2.8 million years of Arctic climate change from Lake El'gygytgyn, NE Russia. *Science*, 337(6092): 315–20.

Meng, Q.-J., Latif, M., Park, W., Keenlyside, N., Semenov, V., and Martin, T. 2012. Twentieth century Walker circulation change: Data analysis and model experiments. *Clim. Dynam.*, 38: 1757–73.

Mildrexler, D. J., Zhao, M. S., and Running, S. W. 2011. Satellite finds highest land surface skin temperature on Earth. *Bull. Amer. Met. Soc.*, 92(7): 855–60.

Miller, G. H., Geirsdóttir, A., Zhong, Y., *et al*. 2012. Abrupt onset of the Little Ice Age triggered by volcanism and sustained by sea-ice/ocean feedbacks. *Geophys. Res. Lett.*, 39: L02708.

Mo, K. C. and White, G. H. 1985. Teleconnection patterns in the Southern Hemisphere. *Mon. Wea. Rev.*, 113: 22–37.

Moses, T., Kiladis, G. N., Diaz, H. F., and Barry, R. G. 1987. Characteristics and frequency of reversals in mean sea level pressure in the North Atlantic sector and their relationship to long-term temperature trends. *J. Climatol.*, 7: 13–30.

Neiman, P. J. and Shapiro, M.A, 1993. The life cycle of an extratropical marine cyclone. Part I: Frontal-cyclone evolution and thermodynamic air–sea interactions. *Mon. Wea. Rev.*, 121: 2174.

Newell, R. E. and Zhu, Y. 1994. Tropospheric rivers: A one-year record and a possible application to ice core data. *Geophys. Res. Lett.*, 21: 113–16.

Oke, T. R. 1978. *Boundary Layer Climates*. London: Methuen. 272 pp.

Oliver, J. E. (ed.) 2005. *Encyclopedia of World Climatology*. New York: Springer. 854 pp.

Osczevski, R. and Bluestein, M. 2005. The new Wind Chill Equivalent Temperature chart. *Bull. Amer. Met. Soc.*, 86: 1453–8.

Palmén, E. 1951. The role of atmospheric disturbances in the general circulation. *Quart. J. Roy. Met. Soc.*, 77: 337–54.

Parthasarathy, B., Munot, A. A., and Kothawale, D. R. 1994. All-India monthly and seasonal rainfall series: 1871–1993. *Theoret. Appl. Climatol.*, 9: 217–24.

Peel, M. C., Finlayson, B. L., and McMahon, T. A. 2007. Updated world map of the Köppen-Geiger climate classification. *Hydrol. Earth Syst. Sci.*, 11: 1633–44.

Penman, H. L. 1963. *Vegetation and Hydrology*. Farnham Royal: Commonwealth Agricultural Bureaux.

Perry, A. H. and Barry, R. G. 1973. Recent temperature changes due to changes in the frequency and average temperature of weather types over the British Isles. *Meteorol. Mag.*, 102: 73–82.

Peterson, T. C., Stott, P. A., and Herring, S. 2012. Explaining extreme events of 2011 from a climate perspective. *Bull. Amer. Met. Soc.*, 93(7): 1041–67.

Philipp, A., Bartholy, J., Beck, C., *et al.* 2010. Cost733cat: A database of weather and circulation type classifications. *Phys. Chem. Earth*, 35: 36–73.

Plaut, G., Ghil, M., and Vautard, R. 1995. Interannual and interdecadal variability in 335 years of Central England temperatures. *Science*, 268: 710–13.

Pokhrel, Y. N., Yeh, P. J., Oki, T., Yamada, T. J., Hanasaki, N., Kanae, S. 2012. Model estimates of sea-level change due to anthropogenic impacts on terrestrial water storage. *Nature Geosci.*, 5: 389–92.

Ralph, F., Neiman, P., and Wick, G. 2004. Satellite and CALJET aircraft observations of atmospheric rivers over the eastern North Pacific Ocean during the winter of 1997/98. *Mon. Wea. Rev.*, 132: 1721–45.

Ramirez-Villegas, J. and Challinor, A. 2012. Assessing relevant climate data for agricultural applications. *Agric. Forest Met.*, 161: 26–45.

Ratcliffe, R. A. S. and Murray, R. 1970. New lag-associations between North Atlantic sea temperature and European pressure applied to long-range weather forecasting. *Quart. J. Roy. Met. Soc.*, 96: 1226–46

Raymo, M. E. and Mitrovica, J. X. 2012. Collapse of polar ice sheets during the stage 11 interglacial. *Nature*, 483: 453–6.

Renssen, H., Seppä, H., Crosta, X., Goosse, H., and Roche, D. M. 2012. Global characterization of the Holocene Thermal Maximum. *Quatern. Sci. Rev.*, 48: 7–19.

Riehl, H. 1962. *Jet Streams of the Atmosphere*. Fort Collins, CO: Colorado State University.

Robinson, A., Calov, R., and Ganopolski, A. 2012. Multistability and critical thresholds of the Greenland ice sheet. *Nat. Clim. Change*, 2: 429–32.

Roche, M. A., Bourges, J. C., and Mattos, R. 1992. Climatology and hydrology of the Lake Titicaca basin. In: Dejoux, C and Iltis, A. (eds.), *Lake Titicaca: A Synthesis of Limnological Knowledge*. Dordrecht: Kluwer Academic Press, pp. 63–83.

Rodda, J. C. and Dixon, H. 2012. Rainfall measurement revisited. *Weather*, 66(5): 131–6.

Rowlands, D. J., Frame, D., Aina, T., *et al.* 2012. Broad range of 2050 warming from an observationally constrained large climate model ensemble. *Nat. Geosci*, 5(4): 256–60.

Ryu, Y.-H. and Baik, J.-J. 2012. Quantitative analysis of factors contributing to urban heat island intensity. *J. Appl. Met. Climatol.*, 54: 842–54.

Satyamurty, P., Nobre, C. A., and Silva Dias, P. L. 1998. The climate of South America. In: Karoly, D. J. and Vincent, D. G. (eds.), *Meteorology of the Southern Hemisphere*. Boston, MA: American Meteorological Society, pp. 119–39.

Sellers, P. J., Hall, F., Kelly, R., *et al.* 1997. BOREAS in 1997: Experiment overview, scientific results, and future directions. *J. Geophys. Res.*, 102(D24): 28731–69.

Seneviratne, S. I., Corti, T., Davin, E., *et al.* 2010. Investigating soil moisture–climate interactions in a changing climate: A review. *Earth Sci. Rev.*, 99: 129–61.

Serreze, M. C. and Barry, R. G. 2011. Processes and impacts of Arctic amplification: A research synthesis. *Global Planet. Change*, 77: 85–96.

Serreze, M. C., Lynch, A. H., and Clark, M. P. 2010. The Arctic frontal zone as seen in the NCEP–NCAR reanalysis. *J. Climate*, 14: 1550–67.

Shakun, J. D., *et al.* 2012. Global warming preceded by increasing carbon dioxide concentration during the last deglaciation. *Nature*, 484, DOI:10.1038/nature10915.

Sheridan, S. C. 2002. The redevelopment of a weather-type classification scheme for North America. *Int. J. Climatol.*, 23: 51–68.

Sheridan, S. C. and Lee, C. C. 2010. Synoptic climatology and the General Circulation Model. *Progr. Phys. Geog.*, 34: 101–9.

Sheridan, S. C. and Lee, C. C. 2012. Synoptic climatology and the analysis of atmospheric teleconnections. *Progr. Phys. Geog.*, 36: 548–56.

Silva Dias, M. A. F., Rutledge, S., Kabat, P., *et al.* 2002. Cloud and rain processes in a biosphere–atmosphere interaction context in the Amazon region. *J. Geophys. Res.*, 107 (D20): 8072.

Simard, M., Pinto, N., Fisher, J. B., and Baccini, A. 2011. Mapping forest canopy height globally with spaceborne lidar. *J. Geophys. Res.*, 116: G04021.

Slatyer, R. O. and McIlroy, I. C. 1961. *Practical Microclimatology*. Paris: UNESCO. 286 pp.

Slonosky, V. C., Jones, P. D., and Davies, T. D. 2000. Variability of the surface atmospheric circulation over Europe, 1774–1995. *Int. J. Climatol.*, 20: 1875–97.

Solomon, S., Qin, D.-H., Manning, M., *et al.* (eds.) 2007. *Climate Change 2007: The Physical Science Basis. Contribution of Working Group I to the Fourth Assessment Report of the Intergovernmental Panel on Climate Change*. Cambridge: Cambridge University Press. 996 pp.

Stevens, B. and Bony, S. 2013. Water in the atmosphere. *Physics Today*, 66(6): 29–34.

Stewart, I. D. and Oke, T. R. 2012. Local climate zones for urban temperature studies. *Bull. Amer. Meteor. Soc.*, 93: 1879–900.

Stolle, H. J. and Bryson, R. 1975. *Climatic Change and the Gulf Stream*. Madison, WI: Department of Meteorology, University of Wisconsin.

Strangeways, I. 2003. *Measuring the Natural Environment*. 2nd edn. Cambridge: Cambridge University Press, 534 pp.

Strauss, R. F. 2007. An international annotated bibliography of climate classifications. http://paws.wcu.edu/strauss/annotatedbib.pdf

Terjung, W. H. 1968. World patterns of the distribution of the monthly comfort index. *Int. J. Biomet.*, 12: 119–51.

Teuling, A. J., Seneviratne, S. I., Stöckli, R., *et al.* 2010. Contrasting response of European forest and grassland energy exchange to heatwaves. *Nat. Geosci.*, 3: 722–27.

Thompson, D. W. J. and Wallace, J. M. 2001. Regional climate impacts of the Northern Hemisphere Annular Mode and associated climate trends. *Science*, 293(5527): 85–9.

Thornthwaite, C. W. 1948. An approach toward a rational classification of climate. *Geog. Rev.*, 38: 55–94.

Thornthwaite, C. W. and Mather, J. R. 1955. The water balance. *Publ. Climatol.*, 8(1): 1–104.

Trenberth, K. E. 2013. The new normal. *GEWEX News, World Climate Research Programme*, 23(1): 2–3.

Trenberth, K. E. and Fasullo, J. T. 2012. Atmospheric moisture transport from ocean to land in reanalyses. *GEWEX News, World Climate Research Programme*, 22(1): 8–10.

Trenbeth, K. E. and Shea, D. J. 1987. On the evolution of the Southern Oscillation. *Mon. Wea. Rev.*, 115: 3077–96.

Trewartha, G. T. 1961. *The Earth's Problem Climates*. Madison, WI: University of Wisconsin Press. 334 pp.

Tucker, G. B. and Barry, R. G. 1984. Climate of the North Atlantic Ocean. In van Loon, H. (ed.), *Climates of the Oceans: World Survey of Climatology. Vol. 15*. Amsterdam: Elsevier, pp. 193–262.

Tuller, S. E. 1968. World distribution of mean monthly and annual precipitable water. *Mon. Wea. Rev.*, 96: 785–97.

United States Committee for the Global Atmospheric Research Program. 1975. *Understanding Climatic Change: A Program for Action*. Washington, DC: National Academy of Sciences. 239 pp.

US Arctic Research Commission. 2003. *Climate Change, Permafrost, and Impacts on Civil Infrastructure*. Arlington, VA: US Arctic Research Commission. 72 pp.

van Heerden, J. and Taljaard, J. J. 1998. Meteorology of the tropics: Africa and surrounding waters. In: Karoly, D. J. and Vincent, D. G. (eds.), *Meteorology of the Southern Hemisphere*. Boston, MA: American Meteorological Society, pp. 141–74.

Van Loon, H. 1967. The half-yearly oscillations in middle and high southern latitudes and the coreless winter. *J. Atmos. Sci.*, 24: 472–86.

Van Loon, H. 1984. The Southern Oscillation. Part III: Associations with the trades and with the trough in the westerlies of the South Pacific Ocean. *Mon. Wea. Rev.*, 112: 947–54.

Vera, C., Higgins, W., Amador, J., *et al.* 2006. Toward a unified view of the American monsoon systems. *J. Climate*, 19: 4977–5000.

Vincent, D. G. 1998. Meteorology of the tropics: Pacific Ocean. In Karoly, D. J. and Vincent, D. G. (eds.), *Meteorology of the Southern Hemisphere*. Boston, MA: American Meteorological Society, pp. 101–17.

Waliser, D. E. and Gautier, C. 1993. A satellite-derived climatology of the ITCZ. *J. Climate*, 6: 2162–74.

Walker, G. T. 1924. Correlations in seasonal variations of weather. *Mem. Indian Met. Dept.*, 24: 275–33.

Wallace, J. M. and Gutzler, D. S. 1981. Teleconnections in the geopotential height field during the Northern Hemisphere winter. *Mon. Wea. Rev.*, 109: 784–812.

Wang, K.-C. and Dickinson, R. E. 2012. A review of global terrestrial evapotranspiration: Observation, modeling, climatology, and climatic variability, *Rev. Geophys.*, 50, DOI:10.1029/2011RG000373.

Warner, T. T. 2004. *Desert Meteorology*. Cambridge: Cambridge University Press. 612 pp.

Wienert, U. and Kuttler, W. 2005. The dependence of the urban heat island intensity on latitude: A statistical approach. *Met. Zeit.*, 14: 677–86.

Wendland, W. M. and Bryson, E. A. 1981. Northern Hemisphere airstream regions. *Mon. Wea. Rev.*, 109: 255–70.

Whittaker, L.M. and Horn, L. H. 1984. Northern Hemisphere extratropical cyclone activity for four mid-season months. *J. Climatol.*, 4: 297–310.

Wild, M. 2012. Enlightening global dimming and brightening. *Bull. Amer. Met. Soc.*, 93: 27–37.

Williams, C. A., Schwalm, C., Hasler, N., *et al.* 2012. Climate and vegetation controls on the surface water balance: Synthesis of evapotranspiration measured across a global network of flux towers. *Water Resour. Res.*, 48: W06523.

Wolff, G. and Gleick, P. H. 2002. A soft path for water. In *The World's Water 2002– 2003*. Washington, DC: Island Press, pp. 1–32.

Wu, L.-X., Cai, W., Zhang, L., *et al.* 2012. Enhanced warming over the global subtropical western boundary currents. *Nature Clim. Change*, 2: 161–6.

Wyrtki, K. 1982: The Southern Oscillation, ocean–atmosphere interaction and El Niño. *J. Marine Technol. Soc.*, 16: 3–10.

Wyrtki, K. 1985. *The Global Climate System: A Critical Review of the Climate System 1982–1984*. Geneva: WMO.

Yarnal, B. 1993. *Synoptic Climatology in Environmental Analysis: A Primer*. London: Bellhaven Press. 195 pp.

Zalasiewicz, J., Williams, M., Haywood, A. M., and Ellis, M. 2011. The Anthropocene: A new epoch of geological time? *Phil. Trans. Roy. Soc. A.*, 369: 835–41.

Zhou, S.-T., L'Heureux, M., Weaver, S., and Kumar, A. 2012. A composite study of the MJO influence on the surface air temperature and precipitation over the Continental United States. *Clim. Dynam.*, 38: 1459–71.

Zhu, Y. and Newell, R. E. 1998. A proposed algorithm for moisture fluxes from atmospheric rivers. *Mon. Wea. Rev.*, 126: 725–35.

Index

absolute vorticity 99
active layer 200
adaptation 201, 202
adiabatic 78
aerosol effects 19–20
aerosol optical depth 20
aerosols 19
African Easterly Jet stream (AEJ) 152
agriculture and climate 212–14
air masses 59
air–sea interaction 135–8
aircraft operations 219
airflow types over the British Isles 119
albedo 18
albedo values for natural surfaces 19
Alert 167
Altiplano 155–7
Amazon River basin 157–9
anabatic winds 76
anemometer 50
angular momentum 92–3
Angstrøm, A. 105
Antarctic Circumpolar Current (ACC) 134
Antarctic ice sheet 168
Antarctica 169
Anthropocene 186–91
applied climatology 206–22
Arctic frontal zone 61–2
Arctic haze 45
Arctic Ocean, 132–4; ice-free 200
Arctic sea ice extent 187
Argo floats 129
arithmetic mean 4
Arrhenius, Svante 187, 188
Asian brown cloud 45
astronomical variations 181
Atlantic Meridional Overturning
    Circulation (AMOC) 128
Atlantic Multidecadal Oscillation
    (AMO) 111, 160
atmospheric chemistry 22–3
atmospheric pressure (see pressure)
atmospheric structure 16
axial tilt (obliquity of the ecliptic) 179,
    180

baroclinic zones 62
barotropic 62
basal metabolic rate 87
Baur, Franz 122
Bergen 164
Bergeron, Tor 35, 59
Berlin 163
Bjerknes, J. 105
black body 16
blocking pattern 99
blowing sand 45
blowing snow 45
Bolivian high 155
bora 78
Boreal Ecosystem-Atmosphere Study
    (BOREAS) 167
boreal forest 165–7
Bowen ratio 25
Brezowsky, H. 122
Budyko, M.I. 44

Callendar, George S. 187, 188
carbon dioxide 22; concentrations
    187; removal 203; sequestration 203
Cenozoic era 175–6
Central England Temperatures
    (CET) 187
Chaco low 154
chemical proxies 176
China, southeast 161
China monsoon 152
chinook 78
chlorofluorocarbons (CFCs) 3
Christchurch 163
circulation modes 103–14
cities, moisture effects 84–5
classifying climate 146
climate 2
climate – elements 14; why it matters
    3–4
climate change 173; effects on water
    resources 215–16
climate classification 8–9, 41
climate effects on transportation 219–20
Climate Extremes Index (CEI) 207

climate forecasts 221–2
climate of land areas 138
climate sensitivity 196
climate services 221, 222
climate statistics 4–6
climate variability 173
climatic aspects of vegetation and soils
    209–12
climatic extremes 207–8, 209
climatic types on land 145–69
climatology 7
clothing insulation 87
cloud cover 32
cloud effects 18
cloud formation 32
cloud liquid water content 31
clouds 31–2
coastal desert 147
coefficient of variation 4
cold front 66
cold island 84
cold spells 191
conceptual models of the general
    circulation 95
conduction 25
Congo River basin 159–60
conservation of angular momentum 93
conservation of potential vorticity 98
continental drift 173
continental Polar air 59
continental synoptic classifications 122
continental Tropical air 59, 213
continentality 141
Coriolis force 53
Coupled Model Intercomparison Project
    (CMIP) 196
crop growth 212
crops – suitable conditions 213
cut-off low 100
cyclone tracks 70

daily maximum urban heat island
    intensities 83
Dansgaard–Oeschger oscillations 182
dating the past 184

day-length 138
deaths from heat and cold 88
Defant, A. 94
degree-day index 27
dendrochronology 178
dendroclimatology 178
derecho 72
desertification 149
deserts 146–9
dew 85
dew point temperature 29
disasters 220–1
distance from the ocean 141
disturbance lines 153
diurnal character of tropical rainfall 159
diurnal range of temperature 27
dry adiabatic lapse rate 56, 78
drought 208–9; drought areas 42–4, 191
Dust Bowl 166
dust devils 148
dust storm 148
Dzerdzeevski, B.L. 123

earliest meteorological network 7
Earth's axis 15
Earth's energy balance 92
Earth's orbit 179
East Antarctica, land ice 175
East Asia –West Pacific monsoon 161
East Greenland Current 131
easterly waves 137
eccentricity of Earth's orbit 180
economic and socio-political issues
    202–4
eddy correlation 41
Eemian interglacial 181
elementary circulation mechanisms 123
elements of climate 14
El Niño 106–8
Empirical Orthogonal Functions 117
energy 15–25
energy balance; of Earth 92; of a human,
    86–7
ENSO outlook 222
environmental lapse rate 29
European Project for Ice Coring in
    Antarctica (EPICA) 181
evaporation 40–2
evaporation pans 41
evapotranspiration 41
extratropical cyclones 64

Fenno-Scandinavian ice sheet 184
Ferrel, William 94
Ferrel cells 94
Flohn, Herman 189
flooding, in Colorado 153; the Indus
    basin 208; the Mississippi River 208
foehn 78
fog 45
fog drip 36
forest climate 79–80
Forks, WA 164
Franklin, Benjamin 130
freeze events in Florida 209
frequency distribution 5
frequency of circulation patterns 189
frequency of extreme warm
    events 201
frequency of hot days 191
frontal zones 61–3
frostbite 87
future climate 194–204

general circulation, 91–101; factors 125
general circulation models (GCMs) 125
geoengineering 203
geological timeline 173–5
geomorphological methods 178
geopotential height 49
geostrophic wind 52
Girs, A.A. 122
glacial cycles 154–7, 176
glacial Lake Agassiz 183
global brightening 190
global climate models 195–6
global conveyor belt 128
global dimming 190
global monsoons 154
global warming 201; history of research
    188
Gobi 147
gradient wind 53
Great Lakes, effects 80
greenhouse effect 187
greenhouse gases, 21, 22, 183; history of
    research on 188
Greenland 168
Greenland ice sheet 168
Grosswetterlagen 122
ground-based remote sensing 12
ground water 217
growth rings in trees and corals 177

Gulf Stream 130–1
gusts 51

Hadley, George 94, 95
Hadley cell 94
hail 36–8
Halley, Edmond 95
harmattan 149
haze 45
health hazards 88
heat capacity 27; of water 135
heat fluxes 23
heat index 29, 88
heat island intensity 83
heat low 149
heat stress in cattle 213
heat wave 29, 207–8; in Europe, 207; in
    Russia 207
heavy precipitation events 191
hemispheric synoptic classifications
    122–3
Hess, P. 122
high altitude 88, 150
high plateaus 154–7, 176
high pressure cell over northeast
    Siberia 48
Himalaya 149
history of world climatology 6–9
hoar frost 85
Holocene, 184; "optimum" 184
horizontal moisture flux 57–9
Hotan 147
hot days, frequency 191
human bioclimatology 86–8
humid subtropical climate 160–1
Humboldt Current 134
Hurricane Sandy 194, 221
hurricanes 138
hydrological cycle 214–15
hydrostatic equilibrium 49
hyperthermal (heat) stress 86
hypothermal (cold) stress 86
hypothermia 87
hypoxia 88
Hypsithermal 184

ice-albedo feedback 188
ice-free Arctic Ocean 200
ice plateaus 168–9
ice roads 219
impacts 199–201

incoming solar radiation 18, 183
infrared radiation 21–3
instability lines 158
insurance and climate disasters 220–1
interception of rainfall 79
interglacials 183
Intergovernmental Panel on Climate
      Change (IPCC), 187; Fifth
      Assessment Report 205; Fourth
      Assessment Report 197
Inter-Ocean Convergence Zone (IOCZ) 159
Intertropical Convergence Zone (ITCZ)
      36, 63
Inversion 29, 169

jet stream, 54–5; low-level, 55; polar
      front jet stream 54; subtropical jet
      stream 54; tropical easterly jet
      stream 55
Juliaca 156

Karaganda 165
katabatic winds 76
Keeling, C.D 188
Kelvin scale 16
Kinshasa 159
Köppen, Vladimir 8–9
Kuroshio 132
Kyoto Protocol 202

Labrador Current 131
La Niña 106–8
Lake Baikal 80
lake climate 80–1
lake effect snow 80
Lake Titicaca 157
Lake Victoria 80
Lamb, Hubert 117, 118
Lamb's "weather types" –
      characteristics 119
land breeze 76
Landsberg, Helmut 82
land and sea effects 127–42
large-scale shelter 75
Last Glacial Maximum 181
latent heat 25
latitude, effects on incoming solar
      radiation 138
Laurentide ice sheet 181, 184
Lhasa 155
lidar 12

lightning 32–3
Lincoln 165
Little Ice Age (LIA) 185
local climate 75–8
local winds 76
London's climate 81
Los Angeles 162
low-level jet 55
lysimeter 41

Madden–Julian Oscillation 112–14
major crops – suitable conditions 213
Manaus 157
Manley, Gordon 189
Marine Isotope Stages (MIS) 183
marine stratocumulus 137
maritime Polar air 59
maritime Tropical air 59
maritime west coasts 164
Medieval Warm Period 185
Mediterranean climates 161–2
mega-droughts 185
mercury barometer 48
meridional cells 94
meridional southerly components 51
mesoscale convective systems 71–2
methane 22, 201
Mexican monsoon 153
microclimate 85–6
mid-Cretaceous period 175
mid-latitude steppe and prairie 165–6
Milanković, Milutin 181
milibar 48
Miocene Climatic Optimum 176
Mitchell, Murray 189
mitigation 202–3
moisture 29–46
monsoon depressions 150
monsoons 149–54
Montreal Protocol 3
mountain sickness 88
mountain wind 77
Mt. Rainier snowfall 164
multicell clusters 71

Namias, Jerome 48, 136
Namib Desert 147
natural climate variability 190
net radiation 23
Normalized Difference Vegetation Index
      (NDVI) 209

normal distribution 5
normals 2
North American monsoon 153
North Atlantic Oscillation 109
North Pacific Oscillation 110
Northern Annular Mode 109
Northern Australia 153
Northern Eurasian Earth Science
      Partnership Initiative
      (NEESPI) 167
Northern Hemisphere currents 129–32

obliquity (axial tilt) 179, 180
occluded front 66
occlusion 66
oceans 128–35
ocean characteristics 128–9
orbital eccentricity 179
optical depth 20
oxygen isotope record 176
ozone, 22; hole 3, 22

Pacific Decadal Oscillation 111
Pacific/North America (PNA) pattern
      108
Palmén, Erik 96
Palmer Drought Severity Index (PDSI)
      43, 213
Pampas grassland 161
Panama seaway 176
Pascal 48
past climates 172–91
permafrost, 166, 168,
      200–1; degradation 219
Permo-Carboniferous period 175
planetary waves 98
plant hardiness zones 201
plate tectonics 173
polar amplification 190–1
polar deserts 147
polar front 61
polar front jet stream 54
polar low 136
polar vortex 97
poleward energy transport 92
pollen 177
pollutants, in urban areas 82
pollution 82–2
polynya 133
positive feedback 188
precession of the spring equinox 179

precipitable water content 30–1
precipitation 34–9
precipitation efficiency 42–2
precipitation gauge 35
present weather 67
pressure 48–50
projected changes 196–9
Proterozoic 173
proxy records 176
pyranometer 18

Quaternary Period 176–86

radiation 16
radiational index of dryness 44
radiative cooling 75
radiative forcing 189–90
raindrops 34–5
rainfall, 35; intensity 35
rain shadow 141
range 4
RAWINSONDE 51
reanalysis 118
regional classifications of synoptic
    circulation types 117–22
relative angular momentum 93
relative humidity 30
renewable energy 217–19
return period 214
Riehl, Herbert 137
Representative Concentration Pathways
    (RCPs) 197, 205
Roaring Forties 135
Rome 161
Rossby, C-G 95, 99

Sahara 147, 149
Sahel 160; shift of rainfall regime 208
Sargasso Sea 132
saturated adiabatic lapse rate 56, 78
saturation vapor pressure 30
scenarios, used in the Fourth
    Assessment Report of the IPCC 197
scirocco 149
sea ice 132
sea and land breezes 76, 134
seacoast zones of convergence or
    divergence 76
seasonal outlooks 222
seasons 15–17
sea-level rise 199–200

secondary air mass 59
seeder-feeder mechanism 35
self-organizing maps (SOMs) 120
semi-annual oscillations 110
sequestration of carbon dioxide 203
sensible heat 24
shift of plant species 201
slope orientation and angle 75
snow accumulation, in a forest 79
snow cover 38–9; effects on climate 142
Snowball Earth 173
snowfall 35
sodar 12
soil 209–10; classification 210–12;
    moisture 210; pH 210; temperature
    210; type 85–6; soil type global
    distribution 212
solar constant 15
solar cycle 186
solar energy 217
solar radiation 15–21; management 203
South American monsoon 153
South Asian monsoon 149–50
South Atlantic convergence zone
    (SACZ) 136, 159
South Pacific convergence zone
    (SPCZ) 136
South Pole station 169
Southern Annular Mode 110
Southern Hemisphere currents 134
Southern Hemisphere three wave
    pattern 112
Southern Ocean 134–5
Southern Oscillation 104–6
Spatial Synoptic Classification (SSC)
    120–2
specific humidity 30
speleothems 177
spring snowmelt 200
squall line 71
squalls 51
standard atmosphere 50
standard deviation 4
Starr, V. P. 94
stationary longwaves 96
stationary planetary waves 96
Stevenson screen (weather shelter)
    27, 85
storm frequency 57–9
storm tracks 199
stratosphere 16

sublimation 80, 142
subtropical anticyclone 48
subtropical jet stream 54
Summit station 168
sunshine duration 18
sunshine recorder 18
sunspots 15
supercell storms 71
surface boundary layer 85
surface energy balance 25
surface receipt of solar radiation 20–1
surface roughness 141
surface skin temperature 86
synoptic circulation types 117–22
synoptic climatology, 116–25; modern
    applications, 125
synoptic weather map 67

taiga 165–7
Taklamakan 147
Tamanrasset 147
Teleconnections 105; patterns, 125
temperate lowlands 162–3
temperature 25–9
temperature lapse rate 27
terrestrial boreholes 177
thermal belt 76
thermal wind relationship 53
thickness 53
thickness lows 66
"Third Pole" 155
Thornthwaite, C.W. 41
thunderstorms 71
Tibet (Qinghai-Xizang) plateau 154–5
topoclimate 75
tornadoes 160
total freshwater reserves 214
trace gases 22
trade winds 51
transient eddies 96
Transpolar Drift Stream 132
trend in ice area 133
tropical cyclones, 70–1, 138, 161
Tropical Easterly Jet stream 55
tropical Pacific, warm and cold
    events 108
tropical and subtropical steppe 160
troposphere 16
tropospheric rivers 57–8
tundra 167–8
typhoons 138

United States, central 120; southeast 160
urban climate, 81–5; moisture effects
      84–5
urban heat island 82–4

valley wind 77
Vangengeim, G. Ya. 122
Vangengeim–Girs classification 122–3
vapor content 29
vapor pressure 30
vasoconstriction 86
vasodilation 86
vegetation cover effect on climate 141
vertical air motion 56
visibility 45
volcanic eruptions 186
vorticity 98; absolute 99; relative,
      98; conservation of potential, 98
Vostok station, 169; ice core record 188

Walker circulation 100
Walker, Sir Gilbert 100, 103
warm and cold events in the tropical
      Pacific 108
warm conveyor belt 66
warm front 66
water balance 42
water resources 214–17; effects of
      climate change 215–16
water vapor 30
weather 2, 3
weather analysis 68
weather radar 12
weather satellites 11–12
Wegener, A. 173
West Africa monsoon 152–3
westerlies 51, 93, 96–9
Westerly Wind Bursts (WWBs) 106
wet lowlands 157–60

White, R. M. 94
why climate matters 3–4
why is the sky blue? 18
Wien's Law 16
wind chill index 29, 87
wind power 217–18
wind rose 51
wind speeds inside a
      forest 79
wind velocity 50
winds 50–6

Yakutsk 166
Younger Dryas 183

zonal circulations 100–1
zonal westerly components 51
zonal wind belts 96–100
zonal winds 94